D1091043

VIDEO DISCS
The Technology, the Applications and the Future

by
Efrem Sigel
Mark Schubin
Paul F. Merrill
with Kenneth S. Christie, John Rusche and Alan Horder

Knowledge Industry Publications, Inc.
White Plains, New York

Video Bookshelf

Video Discs: The Technology, the Applications and the Future

Library of Congress Cataloging in Publication Data

Sigel, Efrem.
  Video discs.

  (Video bookshelf)
  Includes index.
  1. Video discs.   I. Schubin, Mark, joint
author.   II. Merrill, Paul, joint author.
III. Title.   IV. Series.
TK6655.V5S57   621.388′33   80-23112
ISBN 0-914236-56-3

Printed in the United States of America

Table of Contents

List of Figures

List of Tables

1

Introduction

by Efrem Sigel

Its appeal is a heady compound of science fact and science fiction. The video disc is a slim, rapidly spinning circle of plastic that brings sound and color pictures to the TV screen. Stack the discs up in an automatic changer and you can have an entire day and night's worth of television without lifting a finger. Add a stereo audio track and you can enjoy the experience of musical performances that combine richness of sound *and* sight. Use the disc as a data storage device, and you can put the information from a bookcase full of books on a single record. (The resulting disc won't be nearly as decorative as books in the living room, however, and you may find it harder to prop your disc player and TV set on your bed covers on a cold winter's night than to turn the pages of a paperback.) Hook the player up to a computer and you can select just the picture frame you want, or even use the device as an ever-patient instructor that never tires of repeating, that quizzes you at the end of each lesson and that moves forward at the pace of the student.

PROMISE OF THE VIDEO DISC

This is the promise of the video disc, a technology that stands at the threshhold of commercial exploitation in the decade of the 1980s. For 10 years before its introduction in North America, the video disc enjoyed the notoriety of an underground cult. Its appeal was heralded in countless articles in the trade and technical press, and with occasional breathless reports in the consumer hobby magazines. Even professionals who had spent their lives in the communications or electronics fields fell under the spell of its promoters. Stripped of all the rhetoric surrounding it, the video disc seemed to offer three features not available on any other audiovisual medium.

First, the raw material out of which discs are manufactured is far less expensive than the film or magnetic tape used to display sound and images, in part because of the miniaturization techniques whereby thousands of tracks of information are laid down on a single disc. Second, the finished disc is so light and compact that it can be carried, mailed or stored far more easily than can other forms. Finally, an optical video disc—one of the two principal types that have been developed—lends itself to rapid and precise retrieval of a single frame of information, which can be displayed on a screen for as long as the user wants. Thus, the video disc becomes both a teaching machine and an information storage and retrieval medium. As such, it offers alternatives not only to all the audiovisual devices but also to microfilm, magnetic (computer) tape and discs, and paper.

The promises surrounding the video disc hark back to those days in the late 19th century when man first perfected a reliable means of mechanically reproducing sound. The Americans Alexander Graham Bell and Thomas Edison, the Frenchman Charles Clos, the Italian Guglielmo Marconi and the Dane, Valdemar Poulsen, with their inventions of the telephone, phonograph, radio and mechanical recording device, discovered how to convert sound into electrical impulses and then back into sound. Edison, never modest about his achievements, knew that in the phonograph he had hit upon a machine that could change the nature of communication. He foresaw it as an instrument of mass education that would put ordinary people in contact with the great teachers and speakers of the world. Its effect on the music and entertainment industries was only dimly foreseen.

The video disc, by contrast, comes not at the beginning but at the end of a century of extraordinary development of mass communications media. Radio, motion pictures, the record industry, television and video tape have all contributed to an explosion in the quantity and reach of entertainment. Radio, TV and the movies are mass entertainment media that reach audiences numbering in the thousands, millions or tens of millions, all watching or listening to the same program at the same time. The record and the video disc, like the book, are much more selective. They reach one person, one group at a time, and thus have the ability to appeal to specialized information tastes.

CONSUMER AND INSTITUTIONAL USES

Much has been made of the appeal of the video disc to consumers who want a low-cost device for playing back television programs at their convenience. But in the consumer market, video discs must find a place vis-a-vis video cassette recorders (VCRs) that are already on their way to becoming established as a consumer item. By mid-1981, an estimated 2.3 million VCRs had been sold in the U.S. The VCR is inherently more versatile than the video disc player, since it can record programs off the air and also, in conjunction with a camera, be used to make original tapes. Thus, the success of the video disc player rests with the ultimate promise of lower cost—a promise that in early 1980 was a long way from being fulfilled.

In the institutional market, however, the disc has a different appeal. Here, the prospect of an audiovisual device that can retrieve a single image in seconds, at the touch of a button, has sparked the imagination of educators, trainers and all who work in communications. The U.S. Department of Health, Education and Welfare saw the possibility that the disc could be used for communications with the deaf, and gave out grants for creating original programs that could be played on the optical disc. The Army and Navy underwrote research into training programs on the disc.

General Motors, looking for a new way to communicate with its 10,000 dealers around the country, also seized on the optical disc for training and product information. And groups at Utah State, Brigham Young University, the University of Nebraska, MIT and other institutions began doing research on new sorts of instructional programs that would take advantage of some of this technology's unique features: the random access ability, the mixture of motion and still frames, and the promising, if largely untested, marriage of the video disc player with the microprocessor to create a new generation of teaching machine.

A strange dichotomy in development of the video disc is thus at work. In the consumer market, there is great economic pressure, born of competition from VCRs, to offer a machine at the lowest possible price—this is the goal of RCA with its SelectaVision player, using capacitive (contact) technology. A price of under $500 is seen as the only way to get consumers to buy en masse. At the same time, those looking at the disc for institutional uses are eyeing its sophisticated features—features that would push the price of a player above that of a VCR (not to mention the high cost of writing and producing programs that use these sophisticated features). Either one of these approaches is wrong, or two different disc standards will emerge, one for consumers and one for institutions, or—and this must be counted as a third possibility—the disc will not catch on at all, because competing technologies are either more versatile or have too much of a head start.

If this happens, the proliferation of disc formats and the resulting confusion among potential customers will undoubtedly be partly to blame. As of 1981 there were three institutional disc players, from DiscoVision Associates, Thomson-CSF and Sony, and by early 1982 at least three different consumer disc systems are scheduled to be on the market: the Philips/MCA version (also offered by Pioneer), the RCA disc and a model developed by Victor Company of Japan (JVC). Since there is no way to make the three technically compatible, the triumph of one in the consumer market will mean the demise of the other two. But even the triumph of one is not assured.

PLAN OF THIS BOOK

The rest of this book is devoted to providing the information that will let readers judge for themselves whether any disc system will in fact catch on. If there is an underlying assumption in our treatment, it is this: applications and economics determine whether a technology becomes widely adopted, or adopted at all. And it is only by thoroughly understanding the applications and their costs that potential users can know how heavily to invest in a new technology, and whether it is worth discarding what they already have in order to embrace the new. Thus, the plan of this book is to lay the foundation for understanding the video disc by describing the technology in some detail, including the different types of systems, who has developed them and who will be manufacturing them (Chapter 2).

Following this groundwork, the major applications will be discussed in turn: consumer (Chapter 3), educational (Chapters 4 and 5), business (Chapter 6) and information storage and retrieval (Chapter 7). This last application, not to be neglected, treats the optical video disc not as a carrier of visual programming, but as an electronic file cabinet either for published information—the equivalent of 100 books can fit on one side of an optical disc—or for information peculiar to one company or educational institution.

Once the applications have been discussed in detail, the book will return to the question of technology, to discuss those technologies that are competitive or complementary to the video disc, such as video recording, videotext, frame-sorting systems, etc. (Chapter 8). The purpose will be to view the video disc in the context of ongoing development of the TV and information storage systems of the 21st century. Finally, we will attempt in a conclusion to summarize the prospects for video discs in various markets, by sketching some of the underlying economic realities that will shape the behavior of potential purchasers (Chapter 9).

The excitement surrounding the video disc reflects the expectations engendered by striking advances in communications technology in the last half of the 20th century; transistors, computers, communications satellites, video recorders and now, video discs. Our penchant for producing—some would say wallowing in—information seems to have found a counterpart in our genius for inventing machines that store it and move it around. It is well to have a perspective that sees the video disc player as another information machine, not as something unique or revolutionary. By tempering the excitement about the disc's potential with sober analysis about its limitations, we may better understand how best to put this new tool to work in rewarding ways.

2

An Overview and History of Video Disc Technologies

by Mark Schubin

Before examining the technical characteristics of the various information storage systems being referred to as video disc systems, it may be useful to formulate an operating definition of such a system.

In general, a video disc system is an information storage system utilizing a spinning, disc-shaped object (the video disc) as its medium. It is designed for the distribution of prerecorded television programming to viewers via ordinary television sets, offering the maximum picture quality that such sets can display. Finally, video discs are designed to be recorded only at disc manufacturing facilities, and not by their users, whether in the home or in a work setting.

This chapter will, however, include technical descriptions of video disc systems that can record discs, that neither spin nor are disc-shaped, that are aimed at markets other than home consumers, and that provide data storage as their first function and offer television programming only as a secondary benefit. Still, the consumer television disc system will be the primary subject of this chapter, and an examination of its evolution may prove useful in evaluating its future. A chronology of major events in this evolution is given in Table 2.1.

A BRIEF HISTORY OF TELEVISION STORAGE TECHNIQUES

While television is the result of a long series of developments beginning in the last quarter of the 19th century and continuing into the third quarter of the 20th century, the three major forms of television storage—discs, magnetic tape and film—all had their origins in the same year: 1927.

Table 2.1   Landmark Events Leading to Video Disc Systems

Year	Event
1925	First television pictures displayed in London by John Logie Baird
1927	Baird uses waxed phonograph discs to record TV images
1951	Bing Crosby Enterprises shows magnetic video tape recorder
1956	Ampex introduces commercial quadruplex recorder for broadcasting
1965	Magnetic Video Recording offers magnetic disc recording system with stop action and instant replay feature
1975	TelDec's TeD system goes on sale in Germany
1978	Magnavox (North American Philips subsidiary) introduces its laser video disc player with MCA supplying discs
1979	RCA announces licensing agreement with Zenith, CBS
	IBM enters joint disc venture with MCA
1980	U.S. Pioneer enters consumer disc market
	Matsushita chooses VHD/AHD disc system
1981	RCA introduces video disc player

The First Experiments

In that year, television was a crude process for transmitting images from place to place via wire or radio. Cameras required a huge amount of light and receivers presented dim, tiny images. Two scientists, R.V.L. Hartley and H.E. Ives of Electrical Research Products, felt that both the lighting problem and the dim, tiny image problem might be solved by interposing film at the imaging and display points.

On September 14, 1927, they publicly announced their method of intermediate film television. Scenes would be shot with a film camera, the film would then be projected into a television camera, a receiver would expose the television images onto film at a distant location and the film would then be projected onto a screen. Systems were eventually developed that would allow a delay of only a few seconds between the original film's exposure and the second film's projection (in fact, portions of the 1936 Berlin Olympics were transmitted in this fashion), but the system's most lasting feature was the recording of television images on film. Until 1956, that was the only commercially available way to record television signals, and even after the introduction of the video tape recorder in that year, film continued to be used for purposes ranging from network time zone delay to archival storage.

A second means of television signal storage was found in a patent application filed by Boris Rtcheouloff on January 4, 1927 in Britain. Rtcheouloff's proposal would have applied the magnetic recording techniques developed by Valdemar Poulsen at the turn of the century to television. In fact, as a metallic ribbon or tape was to have been used for the recording, this was the first proposal for a magnetic video tape recorder. Unfortunately, magnetic recording technology was very primitive at the time and there seems to be no record that Rtcheouloff ever constructed anything.

The third 1927 proposal for video recording was put forth by John Logie Baird, generally considered the first person to achieve recognizable television pictures. Baird's system was called "Phonovision" and, like Hartley, Ives and Rtcheouloff, he, too, applied a 19th century storage technology to television. Rather than choosing film or magnetic recording, however, he chose phonograph records.

At the time he developed "Phonovision," Baird was transmitting television as a series of tones carried via AM radio stations. He was able to use the limited bandwidth of such stations for his transmissions because his pictures were very coarse, having only 30 scanning lines at 12.5 frames per second (by contrast, the present American television system utilizes 525 scanning lines at 30 frames per second). Thus, recording Baird's television signals upon waxed phonograph discs was simply a matter of recording the same audio-like signal which was being transmitted by the radio stations. Nevertheless, 1927 was the year of the first video disc system—in fact, in 1935, in Selfridges in London, Baird sold prerecorded discs under the "Major Radiovision" label for several months, and applied his color television techniques to discs, as well.

The Fifties: Breakthrough

In 1927 the video disc was simply a means of recording television images. It was a consumer item only because consumers (however few they may have been) were the only ones utilizing television. That was soon to change.

After World War II German audio tape recording technology was introduced to the United States. Bing Crosby was instrumental in the proliferation of that technology, as was John Mullin, a member of the U.S. Army Signal Corps team that brought the first tape recorders back from Germany. Utilizing Mullin as a consultant, Ampex, until that time a motor manufacturer, created the first American audio tape recorder.

Longitudinal recorders

On November 11, 1951, the Electronic Division of Bing Crosby Enterprises, led by Mullin, demonstrated the first working magnetic video tape recorder. It was based on audio tape recording principles and, thus, simply passed tape over a multi-track head at 100 inches per second, a rate which was supposed to allow the recording of video information. The recorder performed poorly, but was hailed as a magnificent engineering achievement nonetheless. (An issue of the *Journal of the Society of Motion Picture and Television Engineers* predicted that such a recorder "will eventually allow full color, stereoscopic pictures with stereophonic sound to be recorded on one strip of tape.")

The difficulty of recording video signals on magnetic tape stemmed from the huge bandwidth of a video signal compared to an audio signal. A high frequency audio signal of 15,000 Hz (cycles per second—the highest frequency transmitted by FM radio stations) would magnetize areas about 2 thousandths of an inch apart on an audio recorder operating at even the high speed of 30 inches per second (the speed of the World War II machines). A video signal of 4.2 MHz (million cycles per second—the highest frequency transmitted by U.S. television stations) would require the magnetization of areas roughly 7 millionths of an inch apart on the same recorder. That was impossible in 1951.

Therefore, the Crosby-Mullin team increased the speed to 100 inches per second, divided the signal into 10 parallel tracks (for an effective speed of 1000 inches per second), recorded synchronization and audio on additional tracks, and still suffered from poor pictures and a resolution limit of 1.69 MHz. (In the U.S. television system, there are approximately 80 lines of resolution per picture height per MHz. Thus, in the full bandwidth of U.S. broadcast television, the smallest picture element which may be transmitted is one divided by (80 x 4.2) or 1/336 of the height of the picture; in the Crosby-Mullin 1951 recorder, the maximum detail was 1/135 of the height of the picture.)

Improved versions of the same sort of recorder were soon demonstrated by the British Broadcasting Corporation, General Electric and RCA, the last showing a color television recorder on December 1, 1953 which consumed 360 inches of 1/2-inch tape each second. At that rate, a 17-inch diameter reel of tape would last for only four minutes. All of these machines were referred to as longitudinal video recorders, since the tape simply passed longitudinally over a fixed head.

Quadruplex recorders

At the same time that the first Crosby-Mullin recorder was demonstrated, however, Ampex, newly involved in tape recording, began work on a different sort of video tape recorder. Since the problem with video recording involved the speed at which the tape passed the head, Ampex engineers sought to increase this speed without increasing the speed with which the tape passed through the recorder. They also tried to improve the picture quality by applying FM signal encoding techniques and by reducing the number of octaves of information which had to be recorded to allow for easier equalization. (An octave is the frequency span from a particular frequency to twice that frequency. Since television utilizes frequencies as low as 30 Hz and as high as 4.2 MHz, it spans more than 17 octaves, which may be compared with the eight octaves transmitted by a typical FM radio station. The same video signal, transposed upwards in frequency to cover from 4 MHz to 8 MHz would fit within a single octave, and it is far easier to adjust the frequency response of a one-octave signal than a 17-octave signal. That adjustment process is called equalization.)

Ampex unveiled its new recorder to CBS affiliates on April 14,

1956, the day before the opening of the convention of the National Association of Broadcasters. In the next four days, it wrote $4 million worth of business.

This new machine, called a quadruplex recorder because it utilized four video recording heads, consumed tape at a rate of only 15 inches per second, yet had a tape to head contact speed (writing speed) far in excess of anything previously dreamed of: roughly 1560 inches per second. This was achieved by spinning the heads transversely across the width of the tape.

Helical recorders

Spinning heads were not an entirely new idea, and had been proposed even for audio tape recording. Thus, in 1953, the Shibaura Electric Company of Tokyo (Toshiba) began work on yet another type of spinning head video tape recorder, this time one in which the tape would be wrapped in a helix around a drum containing a horizontally spinning head. The drum would thus record long diagonal tracks across the tape. Rather than diagonal recording, this type became known as helical, describing the tape path. Toshiba's helical product came out in 1960, utilizing 2-inch wide tape and a consumption speed of 15 inches per second, identical to Ampex's quadruplex machine.

The Market in the Early Sixties

This, then, was the way the video recording market looked in the early 1960s: There were quadruplex recorders consuming 30 square inches of tape each second and priced at levels that even television stations found difficult to afford; there were helical video tape recorders consuming the same amount of tape and priced in the same range; and there were longitudinal video tape recorders that simply did not work.

At a rapid rate, companies introduced helical recorders that were smaller and consumed less tape, as they gave up broadcast quality. Two-inch wide tape led to 1/2-inch; 15 inches per second shrank to 10 and even less. Companies offering helical video tape recorders to the new industrial and educational markets included Ampex, Sony, International Video Corporation and Philips (Norelco). Finally, when it seemed as though helical technology was ready to reach the consumer, Sony introduced its CV (consumer video) series of 1/2-inch, 7.5 inches per second recorders, priced at a fraction of the price of the broadcast recorders, but still far more expensive than even a color television set.

The CV series didn't succeed in the consumer market, but it spawned many successors who banded together to introduce the first helical recorder standard, under the auspices of the Electronics Industry Association of Japan (EIAJ). Sony's version of the EIAJ machines became the AV series, now aimed at the audiovisual market of schools.

Meanwhile, Magnetic Video Recording provided a magnetic disc for video recording that provided stop action and instant replay for a CBS football pickup on July 8, 1965. Improved to add slow motion and, later, color, these video disc systems remain the backbone of sports coverage (see Figure 2.1). In this case, the disc spins at 1800 rpm, a figure derived from the frame rate of U.S. television signals (30 frames per second); each revolution of the disc represents a single frame. A series of recording heads travels radially across the disc during recording or playback. Thus, for normal television, the head would move across the disc at 30 tracks per second. For stop motion, the head simply stops while the disc continues to spin, and for slow motion, the head travels at less than 30 tracks per second, while the disc continues to spin at its 1800 rpm rate.

Alas, these broadcast systems were as expensive as their quadruplex recorder counterparts and, even today, have capacities ranging only from a few seconds to a few minutes of motion.

This, then, was the situation facing consumer electronics manufacturers interested in storage technology. Video tape, even in its crudest form, was expensive, both in terms of machine cost and in terms of tape consumption (still 3.75 square inches per second, even in EIAJ machines). Quality was far below that available from off-air pickup, except, of course, in the expensive broadcast quality machines, and tape duplication was a tedious process: duplicating a single, hour-long program required an hour of time on a recorder and a player, not counting rewinding and set-up time. Duplicating 1000 one-hour tapes took an hour on 1000 recorders or 1000 hours on one recorder, or something in between. There was no way to knock out video programs in the same way that audio discs could be pressed. There was,

Figure 2.1   The Ampex HS-100 video "instant replay" disc. Courtesy Ampex Corp.

however, a way to record video, unattractive though it may have seemed at the time, on discs. This was when the present crop of video disc research began.

COMMERCIAL VIDEO DISC SYSTEMS—
THE TELDEC VENTURE

Conceived as a means of introducing prerecorded television programming into the home, video disc systems had to share one characteristic: each had to be capable of replicating programs, easily and inexpensively, onto a storage medium which was, itself, inexpensive. Even today, low-cost materials and replication are the key characteristics of video disc systems.

One obvious means towards these goals was the emulation of phonograph disc systems, which stamp aural information onto plastic discs for a few cents. A joint venture of Telefunken of Germany and Decca of Britain (not to be confused with Decca Records in the United States), TelDec, developed a video disc system called TeD (for Television Disc), based upon phonograph technology. The disc was stamped in plastic and had grooves with bumps in them, just as ordinary phonograph records do, and was played by a needle riding in the grooves. Even the master discs were prepared using modified disc cutting lathes.

There the similarity ended, however. Just as Bing Crosby Enterprises and the others involved in longitudinal video tape ran into difficulties when trying to apply an audio technology to television, so did TelDec. Since video recording requires some 210 times more bandwidth than audio recording, a typical audio disc lasting 20 minutes might have been expected to last only five seconds if speeded up to a video rate. Alternatively, 210 times as many bumps would have to be applied to the grooves.

In fact, the TeD system was speeded up to 1500 rpm (at the European frame rate of 25 frames per second, or 1800 rpm at the U.S. 30 frames per second rate), and more grooves were added. Even so, playing time was only 10 minutes per disc, with approximately 280 grooves per millimeter. Discs were 21cm (about eight inches) in diameter and were flexible to allow binding into periodicals.

TelDec soon found that needles which would be depressed and released by the bumps in the grooves were impractical since at 1500 rpm (or 1800 rpm) they would snap. Instead, it came up with a rigid, skid design. The bumps of each groove would pass under the rounded end of the skid and get compressed. The rear end of the skid was quite sharp, and as each bump passed this edge, it would suddenly be released, imparting a mechanical impulse to the skid. This impulse would be carried to a piezoelectric transducer (a crystal which produces a voltage related to mechanical stresses), which fed electronic circuitry.

It was also found that this compression/release cycle did not work well if the disc was supported on a turntable, so the disc itself was clamped to the shaft of the motor, with no other support for the disc. This, in turn, created other problems, eventually leading to a sophisticated air cushion support for the disc which actually imparted a curved shape to the otherwise flat disc in the area of the pickup, for improved signal stability.

Finally, the bumps were found to be an unreliable means of deducing anything except the presence or absence of a bump (in other words, the size of a bump could not be accurately related to the amplitude of a signal). For this and other reasons related to signal improvement, the video and audio signals contained on the disc were first frequency modulated before recording. (In frequency modulation, a variation of amplitude is converted into a deviation of frequency from a nominal, or center, frequency. The process is similar whether it is used for FM radio transmission, quadruplex video tape recording or video disc encoding. Because the amplitude of the FM signal carries no useful information, it may be reduced to a series of dots and dashes; the presence of a dot or dash might indicate a wave that is above a certain amplitude and the absence of a dot or dash, the duration that a wave is below that amplitude. These dots and dashes of varying length can, in turn, be converted into bumps of varying length. This process of FM encoding is the basis for virtually every video disc system.)

TelDec was introduced in West Germany on March 17, 1975, the first commercially available video disc system (except for Baird's). Sanyo, General Corporation (of Japan) and Nippon Video Systems were licensees who were expected to handle the U.S. market, since Japan and the U.S. share the same television standard. Georg Neumann GmbH manufactured mastering lathes and professional players, and Techno Products developed an industrial version of the

TeD system. In 1980, General came out with a jukebox holding up to 50 TeD discs.

Running time may have been a drawback for the TeD system, but a 12-disc changer was developed that made possible as much as two hours of playing time, with a maximum wait of four seconds between discs—a one-second version was expected as well. The diamond skid was periodically sharpened in the player itself but replacement skids were available for approximately $8 after the skid wore out. It was designed to last 100 to 200 hours.

Discs went for about $4 to $12, the player for $650, and a prototype changer, shown in late 1976, was to have cost slightly more than $800.

Even with the FM signal coding, TeD was not impervious to dirt, fingerprint oils and scratches, and delivery of discs was initially slowed by the need to add protective sleeves to carry the discs. A final problem with the TeD system related to the mastering process: discs had to be mastered at 1/25 of real time, a process which required all programs for the disc to first be transferred to film. Real-time mastering was still in development in 1976.

Given the various economic and technical drawbacks of the TeD system, it is no surprise that the product failed to catch on in Europe.

At the same time that TelDec was investigating its phonograph-type contact system in the late 1960s and early 1970s, many other companies were experimenting with disc technologies. Among them were Matsushita, MCA, Philips, Sony, i/o Metrics and Thomson-CSF. These systems will be discussed in the remainder of this chapter.

OTHER PHONOGRAPH-TYPE CONTACT SYSTEMS— MATSUSHITA'S VISC

The only other video disc system which would have directly emulated phonograph recording was Matsushita's abortive VISC system—announced in the U.S. in early 1979, but withdrawn a year later. VISC (the name is a contraction of video disc), in fact, went even farther towards the emulation of audio discs than did TeD. Rather than being pressed onto flexible plastic, VISC discs were to be formed out of ordinary phonograph disc materials, utilizing ordinary phonograph disc stampers.

VISC discs would have been 30cm (about 12 inches) in diameter, and would have played for 30 minutes on each side in the VISC I (900 rpm) mode and for 60 minutes on each side in the VISC II (450 rpm) mode. In addition, a smaller VISC "single" was contemplated to offer seven minutes per side at 720 rpm. Mastering was to take place at real time. The needle, called a twist stylus, was actually to be moved by the fluctuations in the grooves, rather than compressing them (the lower VISC speeds may have been instrumental in allowing this). The player was to have cost $480 to $600 in Japan, and a one-hour disc was to have cost approximately 20% to 50% more than a comparable audio disc, at the retail level.

Both the TeD and the VISC systems offered a sort of random access, in the same way that phonograph records may be said to be random access—the stylus may be dropped anywhere. In the players of both disc systems, controls were provided for this purpose, somewhat more sophisticated in the VISC system because of its longer playing time. Both systems also offered stereo sound capability. However, since a single FM signal was to be recorded (as in virtually all of the video disc systems), the number and quality of the audio channels could be changed by a simple modification of the playback electronics, without affecting the mechanical design of the system (unlike video tape systems which require separate tracks and heads for audio signals). Thus, until a unit actually reaches the marketplace, including or omitting such features as the manufacturer deems cost effective, all FM video disc systems are capable of stereo or multilingual sound.

VISC was abandoned by Matsushita in favor of the Video High Density (VHD) system developed by the Victor Company of Japan (JVC), in January 1980.

Other than VISC and TeD, none of the proposed or actual systems involves mechanical contact through compression and release of either stylus or disc. Such compression reduces disc and stylus life, though no exact figure can be placed upon either, since quality would continually deteriorate rather than suddenly cut off. Nevertheless, video disc technologies are often divided into contact technologies and non-contact technologies and, of the former, there are other systems besides VISC and TeD. Some of these are called capacitive systems, because they make use of the electrical characteristic called capacity.

CAPACITIVE SYSTEMS

Capacity is a measure of a device's capability to store electrical charge. A capacitor is usually thought of as two parallel plates, one storing positive charges and the other negative charges. The larger the plates and the narrower the distance between them, the greater the capacity. Current can only flow into a capacitor until it has reached its capacity. However, if the direction of the current's flow is reversed just before the capacity is reached, it can continue to flow until the charges stored in the capacitor have been reversed. Clearly, the larger the capacitor, the less often this flow reversal will have to take place. Thus, capacitors are frequency sensitive devices, higher frequencies passing through more easily than lower ones. A capacitor can thus be used to tune a circuit. In fact, most radios use a capacitor to select only the frequency desired.

A video disc can act as one plate of a capacitor, its stylus as the other plate. In such a case, the video disc must be made from a conductive material and must be coated with a non-conductive material (or the stylus must be) so that the plates of the capacitor do not short out. If pits are etched into the surface of the disc, in the same way that bumps were applied to the TeD and VISC discs, the depth of a pit will vary the capacity of the stylus/disc capacitor. This varying capacity can be used to tune an electronic circuit to reconstruct the recorded signal.

This is the system which RCA plans to introduce, as well as the system utilized by JVC in its VHD system, for which Matsushita abandoned VISC.

The RCA System

The RCA SelectaVision disc system was developed in the early 1970s in the company's Princeton (NJ) labs. In 1979 the corporation's chairman Edgar Griffiths announced it would reach the market in first quarter 1981. The system utilizes a grooved disc, like the VISC and TeD systems, but the sole function of the grooves is to guide the stylus over the pitted disc surface. Like the other systems, the RCA system also uses an FM encoding scheme, and is capable of stereo and multilingual sound. The first RCA players will feature only one sound channel, but this is simply an economy measure based on the electronic circuitry needed for detecting and amplifying the second sound channel and the marketers' view of the value of that channel. It is not a deficiency of the system, and units could easily be modified at a later date to add the second channel.

Like VISC II, the capacity of the RCA system is 60 minutes per side, based upon a disc diameter of 12 inches, and, like VISC II, the RCA disc spins at 450 rpm, meaning that in the U.S. there are four television frames played for each disc revolution (see Figure 2.2). Again, like VISC, the RCA disc master is recorded in real time by a piezoelectric cutting stylus (piezoelectric crystals can convert electrical signals into mechanical motion as well as vice versa). Again, similar to other disc systems, the master is processed through various stages until polyvinyl chloride (PVC) discs may be stamped out in the same way that they are stamped out for audio or VISC. The PVC is made conductive in this case, however, by the addition of fine carbon particles to the plastic resin. As with most other disc systems, the RCA discs may either be pressed or injection molded. The original RCA discs were to have been sandwiches of conductive and insulating layers. However, it seems that the latest conductive plastic composition eliminates the need for additional disc coatings beyond a sprayed-on lubricant and a protective caddy (sleeve). (See Figure 2.3.)

There are somewhat more grooves per millimeter on the RCA disc than on TeD (about 394 versus about 280)—roughly 38 RCA grooves would fit into a single LP groove. The pits are as small as half a micron (less than 20 millionths of an inch—smaller than a single wave of greenish light).

The player offers all of the advantages of the VISC and TeD players to the user, including the quasi-random access search feature, and has been designed to be as simple as possible. In fact, one critical area, correction of the signal for eccentricities introduced by poor alignment or other problems, has been taken care of by an ingeniously simple tone arm stretcher, working something like the coil of a speaker to move the arm in or out, depending upon the error signal fed into it.

A possible feature of this disc system, which may not be introduced in the first players, is the ability to provide a form of stop-action or freezing. Since the disc rotates at 450 rpm, there is no way to prevent

Figure 2.2   RCA video disc, 12 inches in diameter, rotates at 450 rpm; one side carries an hour of programming. There are 10,000 grooves to the inch on the disc. Courtesy RCA.

Figure 2.3 A plastic sleeve protects the RCA disc from wear. The sleeve is inserted in the player (right), depositing the disc on a turntable. To remove the disc, the sleeve is re-inserted into the player. Courtesy RCA.

this freezing from showing a sequence of four frames, rather than a single one. Nevertheless, it is possible for the stylus to jump between grooves without excessive wear on the stylus or disc, so at least this four-frame sequence can be frozen. RCA is not pushing this capability, but has mentioned it in response to the vast capabilities of the various optical disc systems.

It would also be possible utilizing the RCA system (and, for that matter, most of the others) to record a huge amount of extremely high quality audio information on a single disc—one figure mentioned was eight hours of octophonic sound—but rights problems would probably make such a disc excessively costly. No such capability will be built into the RCA players expected to come onto the market in 1981.

RCA's announced target price is under $500. Replacement styli, needed for the player after more than 200 hours, are said to be about $10, and disc prices will be more affected by the program than by the technology.

RCA has signed a number of important licensing agreements permitting other companies to manufacture either disc players or discs. Licensees include Toshiba, Sharp and Nippon Electric in Japan, and Zenith and CBS in the U.S. Zenith will make players, while CBS will build a pressing plant to replicate discs in the RCA format.

JVC's VHD System

The VHD system from Victor Company of Japan (JVC), an affiliate of Matsushita, is extremely similar to RCA's capacitive system, the principal differences being that the JVC system does not use grooves for tracking, and that the rotational speed is 900 rpm, or two U.S. television frames per rotation. The system was announced in 1979. The disc size has also been reduced to a 10-inch diameter, while maintaining one hour of playing time per side. This last feature was accomplished by narrowing the track pitch still farther to 1.35 microns (more than 740 grooves per millimeter, versus RCA's 394 and TeD's 280). Besides allowing smaller discs and players, this size reduction allows a savings of 25% on the disc materials. These were never previously considered to be a major cost item, but now, as an oil-based commodity, are getting more expensive daily. Like the RCA and TeD systems, VHD utilizes a protective caddy as well, to reduce dirt, scratches and fingerprints.

Like RCA's, the VHD disc is made of conductive plastic and may be stamped or injection-molded. Masters are recorded in real time on a photosensitive, glass master by means of a laser (RCA has also experimented with optical and electron-beam mastering techniques). The stylus, however, is 10 times larger (in terms of disc contact surface) than RCA's. JVC claims that this reduces wear significantly. An additional signal, used exclusively for stylus tracking, is recorded onto the disc along with, but separately from, the composite FM signal. This guidance signal, in conjunction with a special electromagnetic tracking system, takes the place of the grooves. (See Figures 2.4 and 2.5.)

Because it is grooveless, the disc may be traversed by the stylus in any fashion, freezing a single track (two frames) or traveling at nonstandard speeds for a slow-motion or fast-motion effect. The slow motion would probably be a bit more jerky than on the optical systems because there are two frames, rather than one, per rotation. JVC claims that the system permits visible high-speed search, variable slow, fast or normal speed (all of these in either the forward or reverse directions) and the display of still pictures. The system also allows for true random access to any of 54,000 pairs of frames (rotational tracks) in a matter of seconds. At some future time, in conjunction with a suitable storage mechanism (as will be discussed in a later chapter), the VHD system might also provide flicker-free access to a single frame. Even now, such access might be implemented simply by turning off the electronics system during the undesired frame, though this would produce a flickering image.

Unlike RCA, JVC is considering audio-only recording as an important feature of the system. In fact, it is usually called VHD/AHD to indicate its use as an extremely high quality audio system. Marketing plans are unknown, however, so there may be two different players sold, or AHD may be offered as an option to VHD.

Stylus life is said to be more than 2000 hours; the fact that the stylus is much larger than RCA's presumably accounts for this longer life.

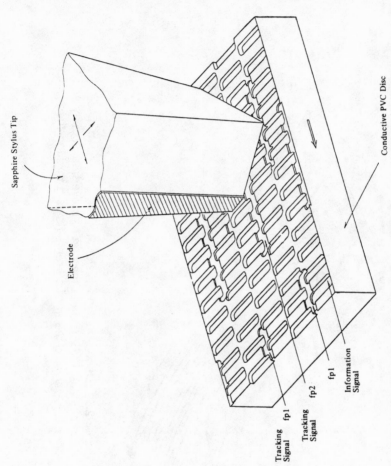

Figure 2.4 VHD/AHD disc from JVC uses microscopic pits that carry tracking signal as well as audio and video information.

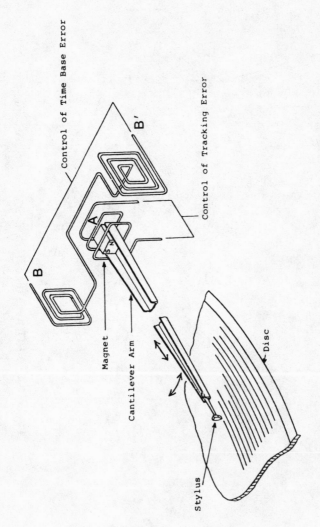

Figure 2.5 Electro-tracking system for the JVC video disc. Three coils (A, B, B′) permit the arm to move transversely and longitudinally, as a result of currents flowing through them.

NON-CONTACT (OPTICAL) SYSTEMS

In addition to such contact technologies as are used in the TeD, VISC, RCA and VHD systems, there are the non-contact technologies, or, as they are sometimes referred to, the optical systems. While it would be possible to create a capacitive system wherein the stylus never touched the disc, it would also be very expensive. Except for a few very singular systems, the optical systems may easily be divided into two categories: those utilizing photographic replication methods, and those utilizing physical (stamping or molding) replication methods. The latter category includes the only video disc systems other than TeD and Baird's which have yet been marketed.

PHYSICAL REPLICATION—PHILIPS AND MCA

By a remarkable coincidence, two organizations, Philips and MCA, demonstrated virtually identical video disc systems in 1972. Both utilized discs which rotated at 1800 rpm (one frame every rotation) and both played the information from the disc without touching it, by utilizing a beam of light. Both used FM coding of a composite signal and both saw physical methods as the appropriate replication means. Just about the only significant difference was Philips' preference for a rigid disc and MCA's for a flexible one.

In January 1976, both organizations, in conjunction with others investigating optical disc systems, issued the first video disc standard, calling for a particular FM coding system featuring two sound channels, a track pitch of 1.6 microns (closer to VHD's 1.35 than to RCA's 2.5 and roughly equivalent to 625 tracks per millimeter), provision for either 12-inch or 8-inch diameter discs, a thickness of 0.2 mm for flexible discs and of 1.1mm for rigid discs, as well as other mechanical and electronic facets.

The discs are mastered with a photosensitive glass mastering process similar to that used by JVC, but once stamped, the discs require additional processes prior to shipment. First a reflective coating and then a protective coating must be applied (see Figure 2.6). The protective coating takes the place of the protective caddy used by the other systems, though Philips/MCA discs are packaged in a sleeve as audio discs are. In limited quantities and in selected markets, both

VIDEO DISC CONSTRUCTION

VIDEO DISC DIMENSIONS

MAGNIFIED VIEW OF A VIDEO DISC

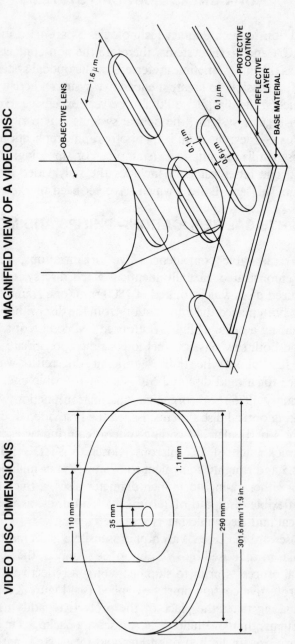

Figure 2.6 Philips/MCA disc measures 11.9 inches in diameter (301.6mm) and is 1.1mm thick. Diagram at right shows the three layers of the disc: a base material, a reflective layer and a protective coating.

disc players and discs have been sold in the United States since the middle of December 1978.

Since the discs rotate at 1800 rpm, these disc systems offer a complete package of features to the user, including all of VHD's features, but adding random access to single frames (rather than frame pairs) and a more steady still frame. Because television frames are composed of two sequential fields, even the Philips/MCA discs with their ability to zero in on a single still frame may present a jittery image if stopped in the middle of rapid action. In addition, since the light source in the optical system is directed at a target beneath the protective surface of the disc, dirt, fingerprints and even scratches have virtually no effect upon the signal quality (and massive dirty areas can simply be washed and scrubbed). Thus, there is no limit yet found on the life of discs in these players, even if a single frame is frozen forever.

The standard discs, unfortunately, offer a playing time of only 30 minutes per side, or, used in the random access mode, 54,000 individual frames per side. To get around this problem, a second form of disc was developed, allowing 60 minutes per side. The rotational speed of these longer-playing discs varies, however, from 1800 rpm at the inner circumference to only 600 rpm at the outer circumference, thereby precluding the use of such features as still frame and slow motion on such discs.

These disc playing systems are laser based, at the moment relying on helium-neon laser tubes. These are glass tubes containing a gas mixture. They may easily be the most expensive single component of the disc player and they are certainly the most fragile. Eventually, these lasers are expected to be replaced by tiny, solid-state laser systems, which should also have a longer life.

In these players, light from the laser is directed through an optical system and focused upon the reflective layer of the disc (see Figure 2.7). The presence or absence of pits in the reflective layer determines the reflection of light back from the layer, which is recaptured by the optical system and sent to a detector. Again, this signal is used to reconstruct the original FM signal, which may then be decoded into its components.

As is the case with all of the disc systems, the price of the disc will be more determined by the program recorded on it than by the materials and the replication process. Nevertheless, the three-stage

PLAYBACK LASER PATH

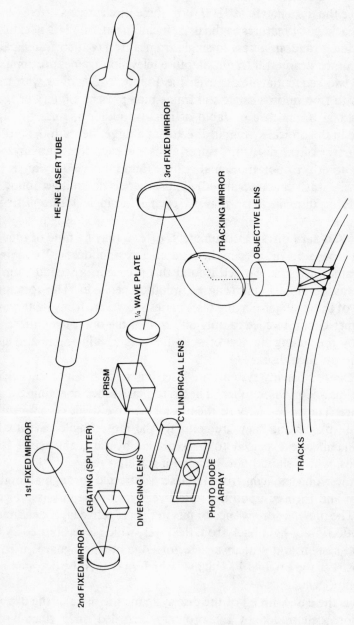

Figure 2.7 Light from a laser is directed through a series of mirrors onto tracks containing audio and video information on the surface of the Philips/MCA disc.

replication process required by these reflective discs (stamping, reflective coating, protective coating) is more complex than that required by any other system. (See Figure 2.8.) Disc movies were priced at $10 to $16 in the first market, but were quickly boosted to $16 to $25. The player itself, targeted at about $500, was actually introduced at $700, was quickly boosted to $775, and has gone through an additional hike since. Life and replacement cost of the laser are purely speculative, but figures as high as 10,000 hours have been used for the lifetime, and figures as high as $100 have been used for replacement cost.

Magnavox, owned by North American Philips, introduced a consumer version of this optical system in December 1978. About the same time, MCA and Pioneer Corp. of Japan, in a joint venture called Universal-Pioneer, began manufacturing an institutional version of the disc player; General Motors ordered some 10,000 units for its dealers. In 1979, IBM joined with MCA in a venture called Disco-Vision Associates to replicate and distribute discs, and possibly to make players as well, with IBM becoming a 25% owner of Universal-Pioneer. Finally, in 1980, Pioneer introduced its own consumer model of a player compatible with the Magnavox model; the Pioneer player went on sale for $750 in a few test markets. (See Figure 2.9.)

Thomson-CSF Disc

The other video disc system which was on the market in 1980—in this case only the educational and industrial market—is that developed by Thomson-CSF. It, too, is an optical, laser-based system, and it, too, utilizes a 1.6 micron track pitch, FM coding, 1800 rpm rotation (for a U.S. version—at the moment, only European versions are available and they utilize 1500 rpm rotation) and most of the other features of the reflective MCA/Philips version just described. However, instead of being a reflective system, Thomson's is a transmissive system, and it utilizes a flexible rather than a rigid disc that measures 12 inches in diameter. The disc carries approximately 30 minutes (or 50,000 frames) on each side. Since it is a transmissive system, though, any of the 100,000 frames can be randomly accessed without flipping over the disc, simply by changing the focal point of the optics from one side to the other. Similarly, the full-hour capacity of both sides of the disc (or the two-hour capacity

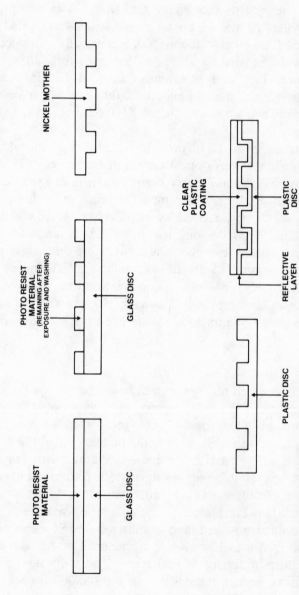

Figure 2.8 Mastering and replication process for the Philips/MCA disc starts with photo-resist material on a glass disc. From this glass disc is made a nickel "mother" disc which is then used to stamp plastic discs. After stamping, discs are covered with a plastic coating.

Figure 2.9 Pioneer optical video disc player is compatible with the Magnavox unit: both will play the same discs. Hand-held unit for remote control is shown below the player.

of the hour-per-side consumer version being developed by Thomson-Brandt) may be viewed continuously without flipping the disc. (Possibly for convenience in using the random access system, Thomson's discs do not have quite a half-hour capability per side, but rather 28 minutes.) The disc is 6 mils (0.15mm) thick.

Mastering is performed with the same laser-etched, photosensitive glass system used by JVC and MCA/Philips, but replication is accomplished in a single step by a stamping or molding process. Since this leaves the disc unprotected, a caddy is used, as in the JVC and RCA systems, and the disc is automatically drawn from the caddy so that it never needs to be touched. The player's optical system is almost identical to that used in the reflective system, except that the light source and the detector are on opposite sides of the disc. It might be argued that either type is slightly better than the other. All of the features of the non-contact, 1800 rpm players are available on Thomson's player as well as some advanced housekeeping functions, such as facilities for computer connection, using the IEEE standard.

The industrial version of this system, the TTV-3620 (see Figure 2.10), went on the market in January 1980, but the price of a consumer version will have to await its development. This unit features all the advantages and disadvantages of the laser-based reflective units. In fact, compatibility of discs has even been demonstrated several times; that is, Thomson discs have been played on a demonstration basis on Philips/MCA players. (In fact, in the laboratory, even a TeD disc has been played on an optical system, and RCA has played its disc by an optical technique.)

Thomson has a number of other manufacturers interested in its disc. Zenith had, for a long time, been a supporter of Thomson's technology, and Minnesota Mining and Manufacturing (3M) announced in early 1980 that it would replicate Thomson discs, while Xerox is planning to use them for data storage.

Sony Disc

Sony has also done development work (announced early in 1978) on laser-based, optical video disc systems, at both 900 and 1800 rpm, with a 1.3 micron pitch (most similar to the VHD), and agreed in September 1979 to exchange patents with Philips. However, on Octo-

Figure 2.10 Thomson-CSF video disc is similar to the Philips/MCA one, but is transmissive, i.e., can be "read" without flipping the disc over, simply by changing the focal point of the optics from one side to the other.

ber 25, 1979, Sony issued the following position statements on video discs:

> "Sony believes that the video disc will not affect the consumer or industrial video tape recorder markets significantly.
> "When the market conditions become suitable, Sony will be ready to take the lead in providing the necessary video disc hardware."

Sony has not only been involved with the development of optical video disc systems, but also with capacitive and with a unique magnetic system, which will be described later.

One final note about the physical replication optical systems is that they, too, are most suitable for use as extremely high fidelity audio storage media. Philips has developed a Compact Disc (CD) only 4.5 inches in diameter for use in a small, laser-based player, with digital audio signals (of the sort that may soon be standardized by the EIAJ). Rotational speed varies from 500 to 215 rpm; track pitch and disc thickness are identical to those used on its video disc. This system uses a solid-state (aluminum gallium arsenide) laser, as it very nearly must because of its small size, and offers one hour of extremely high fidelity (85 dB signal to noise ratio) stereo sound on each side. The prototype player, roughly the size of an audio cassette deck, was first demonstrated in the United States on May 31, 1979. There is no indication that an option to the video disc system might allow playback of these audio discs, but the compatible track pitch and disc thickness lend some credence to that idea.

PHOTOGRAPHIC REPLICATION

The other type of non-contact or optical video disc system uses photographic replication technology, rather than physical replication technology. While a large number of companies have done work in this area (including RCA and Eastman Kodak), two names that stand out are i/o Metrics/Videonics, and ARDEV, a wholly owned subsidiary of Atlantic Richfield.

i/o Metrics

The i/o Metrics system (see Figure 2.11) was first demonstrated early in 1974. It is an extraordinarily simple system. Rather than using FM encoding or any other form of carrier, the i/o Metrics system simply utilizes the raw video signal (with assumed multiplexed audio, in a consumer version) to modulate the intensity of a laser beam onto a sheet of photographic film, in much the same way that an optical sound track for film might be made. As soon as the film is developed by conventional means, the master is completed. To replicate copies, discs of film are placed under the master and a light is turned on. Conventional development yields finished discs. In fact, this contact printing process can be applied to diazo as well as silver-based emulsions, removing the cost of discs from the vagaries of the silver market.

The player spins the disc at 1800 rpm, either aerodynamically or on a glass turntable (both methods were tried). A simple 25-watt light bulb serves as light source and splashes illumination over a wide area of the disc. Microscope optics, in conjunction with an electronic servo tracking mirror, read the varying light intensity. After amplification or modulation, the signal is ready to enter the home TV set, since no decoding is required. Capacity and track pitch were still variable during the last announcement of development, the former in the range of 30 minutes and the latter said to be limited only by the characteristics of the film and of the microscope optics. While such a non-contact, 1800 rpm player should be able to take advantage of the various random access and variable speed features of the other optical systems, i/o Metrics sought, instead, to try for the least expensive player possible. Its white light player with no decoding was to have retailed for $75. A recorder would cost between $15,000 and $20,000, and a duplication system would be as simple as a contact printer, making small quantity runs an economic possibility. Finally, i/o Metrics also experimented with the focus changing possibilities offered by Thomson. At one time, a five-layer disc was proposed, with the change between layers accomplished as simply as changing focus in a microscope. The precise focus was also said to eliminate many of the problems of dirt, fingerprints and scratches. Thus, the

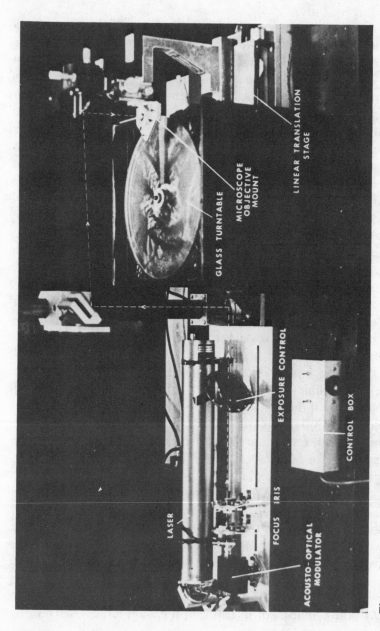

Figure 2.11 i/o Metrics used a technique of modulating the intensity of a laser beam onto a sheet of photographic film to produce a master.

i/o Metrics disc had neither protective coating nor protective caddy.

Subsequently, i/o Metrics went out of business and its technology was never commercially developed.

ARDEV

ARDEV's system, called an Optical Memory Disc System, has its origins in research done by Atlantic Richfield. In fact, the ARDEV Company was established by Atlantic Richfield in February 1978 to demonstrate the technical and commercial feasibility of a photographic video disc system. Some of the goals for the system included real time, economical on-site recording, high interactivity (involving single and multiple track playback), video, audio and/or data recording and playback on the same disc and a non-laser-based playback system.

The disc is nearly 13 inches in diameter, with a wide, 6-micron pitch (almost twice that of even the TeD system). Rotational speed in the U.S. would be 1800 rpm, and there would be approximately 15,000 tracks per disc (8.33 minutes). Duplicates would use diazo, rather than silver emulsions. The player would feature easy computer connection through a built-in microprocessor and both standards of connection (IEEE-488 and RS-232). Time compressed audio would provide at least 13 hours of audio on a disc fully utilized for that purpose (at only 180 rpm). Video, as in the i/o Metrics system, would not be FM encoded. Maximum turnaround time from recording to playback would be 45 minutes. Digitally the disc would store data at speeds up to 10 million bits per second. Access to any bit of data or frame of video would be a maximum of one second. (See Figure 2.12.)

ARDEV was acquired from Atlantic Richfield in June 1981 by McDonnell Douglas Corp., becoming the video disc division of McDonnell Douglas Electronics Co. The purchase price was $2 million "plus royalties on future selected product sales." However, no information is yet available on when, or even if, the system will be marketed. It is obvious, though, that with a bit more than eight minutes per disc and all of the multistorage and computer features, this disc system is aimed at industrial and educational and not entertainment markets. One other capability of the system provides for still pictures accompanied by 30 seconds of audio, something the other optical systems would not seem able to handle.

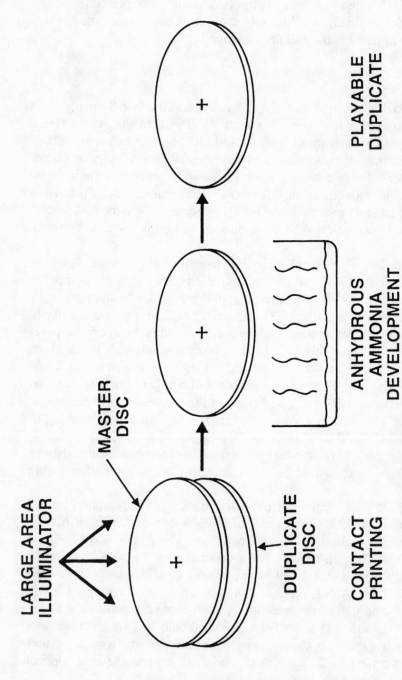

Figure 2.12 ARDEV video disc uses a photographic process of contact printing and development to produce duplicates.

Digital Recording Corp.

Another photographic replication system, but one which may not quite fall into the realm of the video disc, was that developed for Digital Recording Corp. by Battelle's Pacific Northwest Laboratories. This system records raw digital data as dots one micron in diameter on tracks three microns apart. However, the disc is actually a card of photographic material, roughly five inches by seven inches, which is placed in front of a rotating, four-lensed scanner. The scanner spins at 3600 rpm in front of the card, recording curved tracks where the scanner passes. Capacity is said to be 30 minutes of color TV pictures with stereo sound on a 5 x 7 card, but a 5 x 14 card could hold an hour, and so on. The player's cost was estimated at $200 to $300, the materials and replication of a card (which has much the same advantages of the ARDEV and i/o Metrics systems) at $.25. Naturally, as a data storage medium, the system was expected to find wide use as an archival computer memory; one plan would use thousands of film cards, each carrying a billion bits, for a trillion bits of storage or more, with an access time of only a few seconds. As it was described in its most recent reports (five years old), this system had no capability for still frame or variable motion, but was definitely capable of random access.

Hitachi

Two other optical video disc systems deserve brief mention. One was developed by Hitachi and is based around a holographic storage technique. The system utilizes a 12-inch diameter disc and plays for 30 minutes. However, each frame, including sound, is recorded as a single hologram one millimeter in diameter. The disc spins at an extremely slow six revolutions per minute, while a laser reconstructs the hologram for solid state imaging devices. Because of holography's inherent insensitivity to dirt and scratches, no protective caddy is required. Still frames, variable motion and random access are all feasible. Since 1975 when Hitachi announced work on this system, it has also developed reflective system prototypes and magnetic systems.

Optidisc

Another system utilizing a disc which spins at only 6 rpm was developed by French inventor Guy Nathan and is called the Optidisc system. Optidisc is simplest of all the systems in concept: there are tiny pictures recorded on the disc, just as they would be on film. Microscope optics enlarge the images and feed them to a television imaging system (camera tubes, solid-state imaging devices, flying spot photodetectors or some combination). The system was to contain one hour on a single disc, and cost as little as $200 for a player.

Finally, in the optical field, the Japanese Government Agency for Industrial Science and Technology developed a system particularly useful for reflective video disc systems. Called SCOOP (for self-coupled optical pickup), this device contained a tiny solid-state laser and a photodetector in a single package. Utilizing SCOOP, the Japanese Government Agency was able to optically read a TeD disc. More importantly, if 60 SCOOPs were used, coupled by fiber optic links to equally spaced points around a disc, then disc speed might be cut from 1800 rpm to 30 rpm without losing any of the benefits of 1800 rpm rotation (still frame, variable speed, etc.). Of course, in addition to the SCOOPs and the tracking systems needed for them, the electronics package would get more expensive. Another improvement to either optical or capacitive systems would utilize pickups on both sides of the disc to eliminate the need to flip the disc over (as in some old jukeboxes). The Thomson system can handle both sides at once without two pickups.

RECORDING CAPABILITY

All of the systems discussed thus far lack one capability available on the crudest video tape system: the ability to record. Yet there have been video disc systems aimed at the consumer that do include that capability.

One such system was created by Erich Rabe and developed by Wolfgang Bogen GmbH. It is a straightforward magnetic recording system utilizing a chromium dioxide coated disc. However, at the center of the disc is a grooved inner disc, in which a needle rides to guide the outer magnetic head. The system, called MDR (for Magnetic Disc Recorder), featured a 12-inch diameter disc spinning at 156 rpm in one version (geared up from a 78 rpm phonograph), with a 50-micron pitch (eight times greater than even the ARDEV system), for a maximum recording/playing time of 10 minutes. The pitch was to be reduced, eventually, to 25 microns for a 20-minute playing time. A recorder/player was to cost between $350 and $400, and a blank disc was to cost roughly $6.

Another system, not exactly disc-based, was developed by Sony and was called Mavica (for Magnetic Video Card). Actually, each Mavica caddy contains two sheets of magnetic material, each 176 mm by 226 mm (7 x 9 inches), one for video information and the other for stereo information. When the caddy is inserted in the machine, the two cards are drawn out onto two cylindrical frames surrounding two sets of rotating magnetic heads. Thereafter, the machine functions similarly to a helical video tape recorder, except that the card nature of the medium allows rapid random access as well as the advantages of the best 1800 rpm disc systems (still frame, etc.). Unfortunately, the card contains only 10 minutes of program material, and achieves that through the technique of skip-field recording, which might be said to provide less than desirable quality.

It should, however, be pointed out that in 1974 when Mavica was developed, the maximum packing density of magnetic tape was about 1 million bits per square inch. In April 1980, Sony demonstrated a digital video tape recorder with a packing density of 50 million bits per square inch. Applying that technology to Mavica, and assuming no other problems, the single card could now carry more than eight hours of programming. Of course, there would be other problems. Nevertheless, magnetic technology should not be discounted as a possible medium for video disc applications (some other advances will be discussed further in a later chapter). Incidentally, part of the Mavica system was a sort of magnetic contact printer for mass duplication.

Optical disc systems can also be made to record. RCA has been examining the optical video disc as a broadcast recording medium for some time, particularly in relation to digital recording technology. First exhibited in 1976, RCA's system was said to be capable of recording one hour of color video and sound on a $20 disc, when the latest developments were discussed in February 1980. The price of the recorder would be in the broadcast class, but so would the quality. It might be conceivable that a relatively inexpensive spinoff of a broadcast unit might be made, if the broadcast unit is developed. Right now that possibility is bleak, since the system can record and play but cannot erase, and therefore cannot edit. The cost of inexpensive $20 discs can mount up when they must be thrown out every time a mistake is made.

John Locke and Craig Willis worked two and a half years at the University of Toronto on the development of an optical disc recording system which would not have that problem. They announced the results of their work in 1977 and claim to have come up with an optical system exhibiting all of the 1800 rpm non-contact benefits, which will not only record and play but will also erase. Furthermore, the overall system cost is to be approximately $800, while blank discs sell for $2 and prerecorded discs for $10. Disc capacity is to be one hour. The recording medium might be thermoplastic, a material which can be deformed and will retain its deformity, but will return to its original shape upon proper stimulation. CBS Laboratories and Xerox (among others) have used such materials for short-term storage of pictorial information.

SUMMARY

Video disc technology was developed primarily to supply the consumer or institutional customer with an inexpensive source of stored television programming. Most video disc systems are playback-only systems. Grooved disc technologies, for the most part, only allow disc playback at normal rates. Non-grooved technologies, particularly those that utilize discs rotating at 1800 rpm, can offer random access to any frame, stop action and variable speed motion. All systems that need to be stamped or molded can be replicated in much the same fashion as phonograph discs, except for optical reflective systems, which require two additional coating processes. However, these coated discs are relatively impervious to damage. Photographic replication systems offer easy and economical duplication of very small quantities of discs. Some attempts at creating systems which can record in the home have been made. Table 2.2 presents a technical summary of the video disc systems discussed in this chapter, and Table 2.3 gives a technical comparison of video disc systems.

Table 2.2   Technical Summary of Video Disc Systems

I. CONTACT SYSTEMS

  A. Mechanical Compression and Release

    1. Disc Compression: TeD, by TelDec, a joint venture of Telefunken and British Decca. Disc mastering lathes produced by Messrs. Georg Neumann GmbH. Industrial equipment developed by Techno Products. Jukebox introduced in 1980 by General Corp. Other licensees: Sanyo, Nippon Video Systems.

    2. Stylus Compression: VISC and VISC II, by Matsushita. Abandoned January 1980 in favor of JVC's VHD.

  B. Capacitive

    1. Grooved: SelectaVision by RCA. Licensed to BSR, Clarion, General, Mitsubishi, Nippon Electric, Pioneer, Sharp, Toshiba, Zenith and, for discs, CBS.

    2. Ungrooved: Video High Density (VHD) by JVC. Three joint venture companies for player manufacture, disc manufacture and programming acquisition are being formed by JVC, Matsushita, Thorn/EMI and General Electric.

II. NON-CONTACT (OPTICAL) SYSTEMS

  A. Mechanical Disc Manufacture

    1. Reflective: VLP/DiscoVision, developed separately and jointly by Philips and MCA. Manufactured, thus far, by Magnavox and Universal-Pioneer, with discs manufactured by DiscoVision Associates (a joint venture of MCA and IBM, which, in turn, in a joint venture with Pioneer, owns Universal-Pioneer). Other licensees: Grundig, Sharp, Pioneer, Trio-Kenwood and Sanyo. There has also been a patent exchange with Sony.

    2. Transmissive: Thomson-CSF video disc system, very similar to a system which had been under development at Zenith. 3M is planning to manufacture discs. Xerox is planning to use them for data storage.

  B. Photographic Disc Manufacture: i/o Metrics and ARDEV (separate systems).

III. SPECIAL SYSTEMS

  A. Magnetic

    1. Rotating Disc: MDR, developed by Erich Rabe and Wolfgang Bogen GmbH.

    2. Cards: Mavica, developed by Sony.

  B. Pictorial: Optidisc, developed by Guy Nathan.

  C. Holographic: Developed by Hitachi.

  D. Digital Optical Card: Developed by Digital Recording Corporation/ Battelle Pacific Northwest Laboratories.

  E. Optical Recording

    1. With Erasing Capability: Developed by John Locke and Craig Willis.

    2. Without Erasing Capability: Developed by RCA.

Table 2.3   Technical Comparison of Video Disc Systems

Disc System	Replication[1]	Diameter	Speed[2]	Minutes Playing Time[3]	Pitch[4]	Features[5]
Mechanical						
TeD/Telefunken	Stamping	8¼"	1800	10	3.5	A
VISC/Matsushita	Stamping	12"	900	30	2.5	
VISC II	Stamping	12"	450	60	2.5	
VISC Single	Stamping	8"	720	7		
Capacitive						
SelectaVision/RCA	Stamping	12"	450	60	2.5	B, C, D
VHD/Matsushita	Stamping	10"	900	60	1.35	C, E, F, G
Optical						
VLP/DiscoVision (Philips and MCA)	Stamping[6]	12"	1800	30	1.6	C, E, F, G, M
VILP/DiscoVision extended play	Stamping	12"	1800 to 600	60	1.6	H, M
Thomson	Stamping	12"	1800	28	1.6	A, C, E, F, G, I
i/o Metrics	Photographic	12"	1800	30	2.7	A, C, E, F, G, I, J, M, N
ARDEV/Atlantic Richfield	Photographic	13"	1800	8.3	6.0	A, C, E, F, G, J, M, N
Digital Recording	Photographic	5"x7"	0[7]	30	3.0	A, H, J, M
Locke/Willis	Uncertain	12"	1800	30	2.5	C, E, F, G, K, L, N
RCA Optical[8]	Photographic	12"	1800	5.5	2.5	J, K
Magnetic						
MDR	Uncertain	12"	156	10	50.0	J, K, L, N
Mavica/Sony	Printing	7"x9"	0[7]	10	12.5	A, C, E, F, G, J, K, L
Pictorial						
Optidisc	Photographic	12"	6	60	--	A, C, E, F, G, J, M, N
Holographic						
Hitachi Holo.	Photographic	12"	6	60	--	J, M, N

Table 2.3   Technical Comparison of Video Disc Systems (cont.)

1. All systems which may be stamped may also be molded.
2. Rotational speed in rpm for U.S. television standard (generally 1.2 times the European speed).
3. Minutes per side. ARDEV, Mavica and some versions of i/o Metrics have only one side.
4. Spacing between tracks in microns (25 thousandths of an inch).
5. Unless noted, all of the systems are FM coded, all play color video and stereo sound and all require a protective caddy for the disc.
6. VLP/DiscoVision discs require two additional coating steps after stamping.
7. Neither Sony's nor Digital Recording's cards spin. Instead, Sony uses a rotating magnetic head assembly and Digital a rotating lensed scanner.
8. Data from 1976. The system is now theoretically capable of 60 minutes of recording, and data are digitally encoded.

KEY:
A The disc is flexible and can be bound into a publication.
B RCA's SelectaVision system is capable of stereo sound, but initial players will be mono only.
C The system features a visual search mode for locating a segment.
D Some potential for still picture reproduction utilizing special discs and players.
E Random access to still pictures.
F Variable speed motion in forward and in reverse.
G Stop action capability (somewhat poorer in VHD than in the other systems).
H A limited form of pictorial search mode.
I Both sides of the disc may be accessed without the need to flip it over. The i/o Metrics system was said to be capable of accessing even more than two layers simultaneously, for greatly extended playing time.
J The signal recorded on the disc is not a composite FM-coded signal.
K The end-user's video disc machine is capable of recording, as well as playback.
L In addition to recording, the end-user's machine would also be capable of erasing or reusing the disc.
M No protective caddy is required and the disc is relatively immune to dirt and scratches.
N The sound capabilities of the system are uncertain.

REFERENCES

"Agreement On Optical V-Disc Specs," *Television Digest,* January 19, 1976.

"Birth of the mass videogram audience," "Rapid advances in video design and engineering," and "Complete guide to consumer video," *Television, Journal of the Royal Television Society,* Volume 18, Number 2 (March/April 1980).

"Color Video Disk Player at Sony," *Electronic News,* May 29, 1978.

"Digital recording used by home video player," *Electronic Design,* June 21, 1974.

"The Future for Video Disc Systems," symposium notes, Institute for Graphic Communications conference, Carmel, California, December 1-3, 1975.

"General Corp. Unveils Video Disk Player in Japan," *Electronic News,* June 14, 1976.

"High power avoided by video-disk laser pickup from Hitachi," *Electronics,* August 19, 1976.

"Hitachi simplifies video-disk laser pickup," *Electronics,* September 2, 1976.

"Hitachi video disk uses holograms, spins at 6 rpm," *Electronics,* August 21, 1975.

"Home video-disk system creates a new image on photographic film," *Electronics,* April 4, 1974.

"Industrial Video-Disc System Sold," *Electro-Optical Systems Design,* February, 1977.

Information Display, Volume 12, Number 2 (April 1976), entire issue devoted to video disc technologies.

"Intelligent Videodiscs and Their Applications," paper presented by Professor Nicholas Negroponte, Massachusetts Institute of Technology, at the MIT Industrial Liaison Program Symposium, January 15, 1980.

"Japanese work on self-coupled laser to play stylus-recorded video disks," *Electronics,* March 4, 1976.

"Low-cost electro-optical disc stores clear images," *Electronic Design,* January 4, 1977.

"Low-Cost Video Recording," *Electromechanical Design,* March, 1974.

"Mavica: Sony's New Ace," *Video Systems,* December, 1975.

"MCA-N.V. Philips Team to Test Video Disks in U.S.," *Electronic News,* April 19, 1976.

"Optical Digital Recording Concept Offers Industry Applications," *Computer Design,* July, 1974.

"Opto-Digital Recorder Stores TV Shows," *Digital Design,* June, 1974.

"Optical Video Disc For High Rate Digital Data Recording," paper presented by G.J. Ammon, R.F. Kenville and G.W. Reno at 1977 Electro-Optics/Laser Conference. The authors are from RCA's Advanced Technology Laboratories in Camden, NJ 08101.

"Optical Disc Standards," *Television Digest,* January 26, 1976.

"Optical Video Disc System Offers Fast Retrieval of Color Still Pictures," *Computer Design,* July 1977.

The Performing Arts and the Future of Television, Lincoln Center Media Development Department, January 24, 1975.

"RCA, Philips-MCA Team Ready Video Disk Players," *Electronic News,* March 24, 1975.

RCA Review, Volume 39, Number 1 (March 1978), entire issue devoted to the SelectaVision video disc system.

RCA Review, Volume 39, Number 3 (September 1978), entire issued devoted to optical technologies related to the SelectaVision video disc system.

"RCA video-disc entry gives hour of inexpensive viewing," *Electronic Design,* March 15, 1975.

"A Review of Video Disc Principles," speech by A. Korpel, Zenith Radio Corp., Chicago, at the Symposium of the Society for Information Display, May 21, 1974.

"$75 home video player promised in 15 months," *Electronics,* November 14, 1974.

"Television on a silver platter," *IEEE Spectrum,* August, 1975.

"The $10 laser is coming," *Electronic Products Magazine,* November, 1975.

"Thomson-Brandt readies a color video disk recorder," *Electronics,* May 30, 1974.

"A Transmission Mode Optical Video Disc System," speech by Robert L. Whitman, Zenith Radio Corp., Chicago, at the Symposium of the Society for Information Display, May 21, 1974.

"Two Firms Develop TV Disk Changer," *Electronic News,* November 22, 1976.

"Two video disc systems to be marketed in 1976," *Electronic Design,* April 12, 1975.

Urban Telecommunications Forum, Volume 4, Number 29 (January 1975), entire issue devoted to video discs.

"Varying Capacitance Creates TV Playbacks," *Digital Design,* October, 1975.

"VideoDisc aims at simple player unit," *Electronics,* June 12, 1975.

"The Video Disc: Count on it," *Broadcast Management/Engineering,* March, 1974.

"The Video Disc Looks Ready For The Consumer," *Broadcast Management/Engineering,* May, 1975.

"Videodisc Standards Outlined," *Electro-Optical Systems Design,* March, 1976.

"Video Disc Technology," speech by Joseph Markin, Zenith Radio Corp., Chicago, at the Symposium of the Society for Information Display, May 21, 1974.

"Video Disc Update," *Electro-Optical Systems Design,* May, 1975.

"Video-disk battle goes public," *Electronics,* April 3, 1975.

"Video-disk system from Thomson aimed at commercial uses," *Electronics,* February 19, 1976.

"Video Playback A Step Closer," *Electro-Optical Systems Design,* September, 1974.

"Video Playback Discs: The Poker Game of the Century," *Electro-Optical Systems Design,* January, 1975.

"Video players for the home turning to card and disc units to cut costs," *Electronic Design,* June 21, 1974.

"Video recorder to debut in color," *Electronics,* April 4, 1974.

Manufacturers' literature and press releases from: Matsushita, JVC, U.S. Pioneer Electronics Corp., RCA, Messrs. Georg Neumann GmbH, Panasonic, MCA, N.V. Philips Gloelampenfabriken, ARDEV, Magnavox, DiscoVision Associates, Sony, and Thomson-CSF.

Personal conversations with representatives of ARDEV, RCA, Philips, DiscoVision Associates, Thomson-CSF, Sony, U.S. Pioneer, JVC, Panasonic, i/o Metrics, Arvin/Echo Sciences, Ampex, Zenith, Techno-Products, Neumann and Magnavox.

Also countless trade conferences and exhibitions.

3

The Consumer Market for Video Discs

by Efrem Sigel

The video disc was born of dreams for a vast consumer market ready and willing to buy an inexpensive device to play back recorded TV programs. The dream is still alive, and most of the attention to the disc in the business world and in the press focuses on this lucrative, but elusive, home market.

As subsequent chapters of this book will show, educational and industrial markets may well be a more promising avenue for the disc, or at least for one version of it, the laser-based optical player. Nevertheless, given the enormous economies of scale that are present in making a product for the consumer market, the success or failure of the video disc here will have considerable impact on its future in other markets.

THE EUROPEAN CASE

On the face of it, Europe is a more promising market for the video disc than the U.S. One reason is the scarcity of television programming in Europe. Broadcasting is largely in government hands; some countries have no more than a couple of channels on the air, and stations transmit fewer hours than in the U.S. Moreover, advertisers who would like to be able to use the great selling ability of television are frustrated by limitations or outright bans on advertising.

All this seems ready-made for video cassette recorders or video disc players that put programming under the control of the consumer, open the medium up to advertising and give many more program creators a chance to sell their wares.

The theory is admirable, but the practice is something else. The life of the TeD disc systems in Germany and elsewhere was notable for its brevity. Manufacturing and disc replication problems hurt (these have marked the introduction of the optical disc into the U.S. as well) and so did the inherent limitation of the TeD disc's 10-minute capacity.

Related to this limitation, but a problem in its own right, was the absence of enough programming. Unless programs that consumers want to see are released at the same time a disc player is introduced, the hardware is just another box-like shape taking up space in the living room. The TeD sponsors spent years trying to convince major entertainment and publishing companies to put materials onto TeD discs, but met with only begrudging cooperation. Because the disc is a playback-only device, the cleverness of the engineers building the machine is insignificant compared to the importance of getting the right programming to offer to viewers.

The Telefunken players went on the market in March 1975, but within 12 months, the system was widely acknowledged to have flopped. As *Variety* put it, "Sales were slight and even nil," even during the Christmas season. Probably only a few thousand were sold, and Telefunken suffered huge financial reverses as a result of its misadventure. The system remains alive, however, and various licensees continue to tinker with it, especially in Japan.

THE PHILIPS/MCA CASE

The curtain on act two of the consumer video disc drama went up in Atlanta, GA, in December 1978. Magnavox, the subsidiary of North American Philips—which in turn is controlled by N.V. Philips, the Netherlands—and MCA Inc. of Universal City, CA, had assured the press they would have players and discs on sale by the end of 1978. They met their deadline, though it was largely for show. A few hundred players went on sale in three retail chains in Atlanta for $695 each. They were quickly bought up by a most atypical group of consumers, consisting of hobbyists, popular electronics writers, novelty seekers, "profiteers" (who resold the scarce machines), and representatives of rival manufacturing and entertainment companies.

The first MCA disc catalog listed 202 titles, more than half of them feature films. Universal Pictures, the MCA movie subsidiary, contributed 50 of its films, among them such blockbusters as "Jaws" and "Jaws 2" as well as "Airport '77" and "American Graffiti." All were priced initially at $15.95. Paramount offered "Love Story" while Warner Brothers chipped in with "Deliverance." From Walt Disney came "Kidnapped" and "At Home with Donald Duck." There was also a smattering of how-to and informational programs: cooking lessons with Julia Child, "Gene Littler's Golf" and "Total Fitness in Thirty Minutes a Week." Table 3.1 shows the MCA catalog as of mid-1980.

Throughout 1979, even though the system also went on sale in Seattle and later in Dallas, the supply of both players and discs was uneven. Discs were a particular problem, and it soon became apparent that MCA was having serious problems in its Costa Mesa, CA, duplication plant. Reports circulated that the reject rate on discs coming off the stamping line was 50% or more, a staggering percentage for any manufacturing process. The turnaround time (period between delivery of a completed program to the plant and shipping finished discs) was often measured in months, not weeks or days.

By mid-1981, the laser optical format seemed to have gained new momentum. DiscoVision Associates claimed to have resolved many of its production woes, and the Pioneer replicating plant in Kofu, Japan, jointly owned by DiscoVision and Universal Pioneer, had gone into full production. 3M had announced it would also replicate video discs for the laser optical system late in 1981. Meanwhile, major studios such as Paramount, Fox and Columbia released blockbuster titles in laser optical format.

RCA, MATSUSHITA AND IBM

As the MCA and Philips team limped along in 1979, rivals were busy laying plans for competitive products. RCA moved to steal some of the Magnavox and MCA thunder by announcing in January 1979 that it would definitely enter the disc market. During 1979 and 1980, it announced licenses with such companies as Paramount, Avco Embassy, United Artists, Walt Disney, Viacom, MGM, Rank Film Distributors, ITC and NFL Films.

Specific programs to which RCA acquired rights included: "The Graduate," "Carnal Knowledge," "Godfather I" and "Godfather II," "Hamlet," "Rocky," "Annie Hall," "A Charlie Brown Christmas," "Jesus of Nazareth," "20,000 Leagues Under the Sea," "The Shaggy Dog," "Star Trek" and "Victory at Sea." RCA also licensed musical concerts from Don Kirshner and Elton John, as well as a program featuring Dr. Benjamin Spock on caring for newborn

Table 3.1 MCA DiscoVision Catalog, Mid-1980

Feature Films	Price
Airport '77	$24.95
American Graffiti (Stereo)	24.95
Battlestar Galactica	24.95
The Bingo Long Traveling All-Stars & Motor Co.	24.95
The Birds	24.95
Blue Collar	24.95
Car Wash	24.95
The Choirboys	24.95
Coal Miner's Daughter	24.95
Diary of a Mad Housewife	24.95
Dracula (Stereo)	24.95
Earthquake	24.95
The Eiger Sanction	24.95
The Electric Horseman	24.95
Family Plot	24.95
FM (Stereo)	24.95
Frenzy	24.95
Gray Lady Down	24.95
The Great Waldo Pepper	24.95
The Greek Tycoon	24.95
Heroes	24.95
High Plains Drifter	24.95
Jaws	24.95
Jaws 2	24.95
The Jerk	24.95
Jesus Christ Superstar (Stereo)	24.95
Joe Kidd	24.95
The Last Remake of Beau Geste	24.95
The Last Married Couple in America	24.95
Love Story	24.95
Luther	24.95
MacArthur	24.95
Midway	24.95
National Lampoon's Animal House	24.95
1941	24.95
The Nude Bomb	24.95
The Other Side of the Mountain	24.95
Psycho	24.95
Rollercoaster	24.95
Rooster Cogburn	24.95
Same Time, Next Year	24.95
Saturday Night Fever (Stereo)	24.95
The Seduction of Joe Tynan	24.95
The Seven Per-Cent Solution	24.95
Sgt. Pepper's Lonely Hearts Club Band (Stereo)	24.95
Shenandoah	24.95
Slap Shot	24.95

Feature Films	Price
Slaughterhouse Five	$24.95
Smokey and the Bandit	24.95
The Sting	24.95
Three Days of the Condor	24.95
Thoroughly Modern Millie	24.95
To Kill a Mockingbird	24.95
Which Way Is Up?	24.95

Classic Feature Films

Abbott and Costello Meet Frankenstein	$15.95
Animal Crackers	15.95
The Bride of Frankenstein	15.95
Buck Privates	15.95
Dracula (1931)	15.95
Francis, The Talking Mule	15.95
Frankenstein	15.95
Going My Way	15.95
If I Had a Million	15.95
The Incredible Shrinking Man	15.95
The Lives of a Bengal Lancer	15.95
The Lost Weekend	15.95
Ma and Pa Kettle	15.95
Ruggles of Red Gap	15.95
The World of Abbott and Costello	15.95

Walt Disney Productions

Kidnapped	$24.95
Adventures of Chip 'n' Dale	9.95
At Home With Donald Duck	9.95
Coyote's Lament	9.95
Kids Is Kids	9.95
On Vacation with Mickey Mouse & Friends	9.95

Television Movies

The Bionic Woman	$ 9.95
Cyborg: The Six Million Dollar Man	9.95
Duel	15.95
The Hardy Boys: Mystery of the Haunted House	9.95
Tom Sawyer	15.95

Non-Feature Films (Home)

Julia Child: The French Chef The Omelette Show	$ 5.95
Quiche Lorraine & Co.	5.95
To Roast a Chicken	5.95
The Romagnolis' Table Abruzzi Specialties	5.95
Made in Milan	5.95
A Roman Family Dinner	5.95
Greek Cooking With Theonie Baklava/Orange Sweets	5.95

Table 3.1 MCA DiscoVision Catalog, Mid-1980 (cont.)

Non-Feature Films (Home)	Price
Moussaka/Baked Spaghetti	$5.95
Spinach Pie/Dolmathes	5.95
Needlecraft with Erica	
Satin Stitch Chains	5.95

Sports—How To

Better Tennis in Thirty	
Minutes	$5.95
Gene Littler's Golf	9.95
If You Can Walk/Listen	
to the Mountains	9.95
Swimming: Freestyle	
and Backstroke	5.95
Swimming: Breast Stroke	
and Butterfly	5.95

Sports—Spectator

The Big Fights: Ali vs.	
Folley and Williams	$5.95
Louis vs. Conn	
1st and 2nd Fights	5.95
Marciano vs. Wolcott	
and Moore	5.95
Robinson vs. Graziano	
and La Motta	5.95
NFL Films	
Catch It If You Can	5.95
The Gamebreakers	5.95
The Runners	5.95
They Call It Pro Football	5.95
Trials and Triumphs	5.95
Young, Old and Bold	5.95
The Magic Rolling Board/	
Skateboard Safety	5.95
Mammoth Mountain Adventure	5.95
The Moebus Flip	5.95
Sentinel: The West Face	5.95
Ski Racer	5.95
Winterwings	5.95

Informational

The Undersea World of	
Jacques Cousteau	
Sleeping Sharks of Yucatan	$9.95
The Singing Whales	9.95
The Sound of Dolphins	9.95
Tragedy of the Red Salmon	9.95
Octopus, Octopus	9.95
Coral Divers of Corsica	9.95
Smile of the Walrus	9.95
Unsinkable Sea Otter	9.95
Jane Goodall: The World of	
Animal Behavior	

Informational	Price
The Baboons of Gombe	$9.95
Wild Dogs of Africa	9.95
The Hyena Story	9.95
Lions of Serengheti	9.95
World at War	
Bonzai	9.95
Morning (D-Day)	9.95
Genocide	9.95
The Bomb	9.95

Educational

CPR/Choking	$5.95
Math That Counts	9.95
The Solar System/	
The Universe	9.95
What Makes Rain?/Storms	9.95
Silent Safari	9.95
Money in the Marketplace/	
Choosing What to Buy	9.95
VD: The Hidden Epidemic	5.95
Archaeological Dating/	
The Big Dig	9.95

Self-Improvement

Smoking: How to Stop	$5.95
Total Fitness in Thirty	
Minutes a Week	5.95

Religion and Moral Values

The Guide	$5.95
The Way Home	5.95
The Making of a Torah/Portrait	
of a Jewish Marriage	9.95
Who Is God? Where Is God?/	
God's World, Our World	5.95
Forgive and Forget/	
Thank You, Thank You	5.95

The Arts

National Gallery:	
Art Awareness Collection	$9.95
The Bolero	5.95
The Art Conservator	5.95
Le Corbusier	5.95

Music

ABBA (Stereo)	$19.95
Elton John at Edinburgh	19.95
Loretta (Stereo)	19.95
Olivia	19.95

babies (see Table 3.2).

Nor were the Japanese companies silent. Matsushita caused a stir at a meeting of the International Tape Association, an industry group, in spring 1979 when it showed a prototype of its VISC-O-PAC disc, supposedly smaller and less expensive than the RCA disc. Later, Matsushita abandoned VISC-O-PAC in favor of the VHD/AHD system developed by Victor Company of Japan (JVC), its affiliated firm.

But the most dramatic announcement of 1979 came neither from RCA nor the Japanese. It came instead from Armonk, NY, home of the world's largest computer company. In September, IBM stunned the entertainment and consumer electronics industries by announcing that it was joining forces with MCA in replication of video discs. IBM became a 50% partner in DiscoVision Associates, a firm whose chief assets were MCA and IBM disc patents, the replicating plant in Costa Mesa and a substantial amount of cash provided by IBM.

The wording of the announcement left ambiguous whether the firm would pursue institutional as well as consumer applications. But it was undeniable that IBM's computer expertise would give it both technological and marketing strengths in developing the so-called "smart" video disc player that incorporates a microprocessor for sophisticated educational applications.

The U.S. video disc market became a battleground in earnest by year-end as RCA announced in December 1979 that it would introduce its SelectaVision player by the first quarter of 1981. Licenses announced with CBS for disc pressing and with Zenith for player manufacture gave RCA a powerful boost in 1980, as did its success in getting the giant retail chains, Sears Roebuck, J.C. Penney and Montgomery Ward, to commit to selling its disc system. Hitachi, Sanyo and Toshiba also signed up as RCA licensees in the U.S. By January 1981, manufacturers accounting for almost 60% of color television sales in the U.S. were committed to the RCA capacitance electronic disc (CED) system.

Meanwhile, the MCA/Philips/IBM camp gained another adherent when Pioneer Electronic Corp. of Japan introduced an optical player in the U.S. in June 1980. Pioneer was already a partner in Universal-Pioneer, which was manufacturing the industrial version of the DiscoVision player and which had turned out 10,000 such players for General Motors and its dealers.

Matsushita and JVC reached agreements with General Electric

Table 3.2   Sample RCA SelectaVision Disc Programs—Licenses Announced in 1979 and 1980

Avco Embassy Pictures
The Graduate
Carnal Knowledge
Day of the Dolphin
The Lion in Winter
The Producers

Filmverhuurkantoor de Dam B.V.
The Kid
The Gold Rush
City Lights
The Great Dictator
Limelight
Modern Times

ITC Entertainment, Ltd.
Jesus of Nazareth
To Russia . . . With Elton
  (Co-licensed by
  Black Lion Films, Ltd.)

The Jackson Co.
Caring for Your Newborn—
  Dr. Benjamin Spock Shows
  You How

NBC
Victory at Sea
Heidi
Hans Brinker
The Louvre
Tut: The Boy King

NFL Films, Inc.
Super Bowls
NFL Games

Paramount
Star Trek (10 episodes)
Grease
Saturday Night Fever
Godfather I & II
True Grit
Heaven Can Wait
Foul Play
The Longest Yard
Bad News Bears
Chinatown
Love Story
The Shootist
Rosemary's Baby
Romeo and Juliet
Gunfight at the OK Corral
Shane
Stalag 17
Sunset Boulevard
Serpico
The Ten Commandments
Star Trek—
  The Motion Picture
Escape from Alcatraz
Starting Over
American Gigolo
Nijinsky
North Dallas Forty

Rank Film Distributors
Henry V
Hamlet
Red Shoes
Odd Man Out

This Sporting Life
Great Expectations
Oliver Twist
39 Steps
The Lady Vanishes

United Artists Corp.
Rocky
Coming Home
Annie Hall
Hair
Fiddler on the Roof
West Side Story
Exodus
Semi-Tough
Pink Panther (all)
In the Heat of the Night
The Great Escape
The Apartment
Hospital
Some Like It Hot
Marty
Judgement at Nuremberg
The Alamo
The Magnificent Seven
Paths of Glory
Elmer Gantry
Casablanca
The Maltese Falcon
Treasure of Sierra Madre
Key Largo
Yankee Doodle Dandy
Red River
Little Caesar
Sergeant York

United Features Syndicate and Lee Mendelson & Bill Melendez Productions
A Charlie Brown Christmas
A Charlie Brown Thanksgiving
It's The Great Pumpkin
  Charlie Brown
(Plus 18 other Charlie Brown
  titles)

Viacom Enterprises
The African Queen
Terrytoon Cartoon Library
Roustabout
Blue Hawaii
Fun in Acapulco
Girls, Girls, Girls
G.I. Blues
Paradise Hawaiian Style
King Creole

Walt Disney Productions
20,000 Leagues Under the Sea
The Love Bug
The Absent Minded Professor
The Shaggy Dog
The Prince and the Pauper
Kidnapped
Escape to Witch Mountain

in the U.S. and Thorn EMI in Britain to manufacture its VHD/AHD system.

CONSUMER DISC ECONOMICS

Nothing is more complex than setting a price for a new piece of consumer equipment. To the question "How much will it cost?" the production manager can turn around and ask, "How many do you want?" Costs that would seem astronomical if only a few thousand units were made drop sharply when the production runs climb into the hundreds of thousands. Conversely, costs that are perfectly reasonable in the context of mass production runs may become unsupportable when actual production and sales volume turn out to be much smaller.

This Catch-22 has impaled manufacturers of video disc players on a very pointy dilemma. Setting the price of the player too high risks scaring away the very consumers that can make possible a mass market. On the other hand, setting the price too low means sacrificing revenues that could be used for marketing, promotion, distribution and the education of customers to the existence of a new product.

The dilemma is by no means academic. For years the price of $500 was discussed as the goal of a consumer disc player. Yet when Magnavox's Magnavision machine was finally offered to the public in December 1978, the retail price was $695, not $500. Its subsequent history was the reverse of what might be expected. Instead of the price dropping as the company developed the market, the price went up—in spring 1979, it was increased to $775. MCA did the same thing with the discs. They were originally priced at $15.95 for recent movies, yet in May 1979 the price was jacked up to $24.95. Even at that price, it's easy to surmise that the pressing and distribution of discs had to be a money-losing business, given the small quantities involved.

In mid-1980, Pioneer introduced its own optical player, fully compatible with the Magnavox model, at a price of $750. The Pioneer machine incorporated capabilities not available on the Magnavox unit, like a push-button search feature that enables the viewer to locate the exact picture frame desired. Thus the consumer is actually getting more for less money with the Pioneer player than with the Magnavox player.

Still, the price of $750 is about what consumers would pay for a discounted video cassette recorder in 1980. And, of course, a VCR can record programs off the air, as well as play them back. Retailers have been expecting VCR manufacturers to offer a basic VCR for a suggested retail price of $595 by early 1981. It is therefore realistic to expect that with discounting, some VCRs will be available for close to $500. Not only is this price substantially less than that of the Magnavox and Pioneer players, but it's precisely the announced target of the RCA player that was due on the market by early 1981.

WHAT DO CONSUMERS WANT?

Gauging the success of video disc players in the consumer market involves answering two questions: 1) Are consumers interested in the features of the video disc at any price? 2) Are consumers interested enough in the added features of the optical players to pay a premium for this particular technology?

Question number one could be rephrased as follows: if a VCR were available for the same price as a video disc player, and if prerecorded tapes sold for as little as discs, would there be any reason to buy disc players?

Surely the answer to this question must be "no." Only the more esoteric features of the optical disc system—random access, freeze frame—would then justify its purchase, and only hobbyists intent on very intensive use of a disc player would be willing to pay the price.

Disc vs. Tape

No one has more experience in dealing with video disc consumers than Rich's, the Atlanta-based retail chain where the Magnavision player first went on sale. At one outlet, people waited all night to buy the 30 available players, says Rich's TV buyer, Hank Freedman, but the sales pace slowed considerably after the first months. Freedman described a lot of the early buyers as "profiteers" who bought the scarce players and then re-sold them at a profit. His description of the disc player owner runs something like this: "he has a VCR—in 50% of the cases; he knows the difference between VCR and disc; he wants to add to his collection of sophisticated home electronic equipment; he's affluent—usually a college graduate with an income of $20,000 to $25,000."

Rich's sells VCRs, too, and finds that when a customer can only buy one item, it's the VCR that is chosen. According to Freedman, this is because a "VCR is something a person is in the market for, whereas a lot of disc sales are impulse." The VCR's ability to record is the decisive feature in motivating the consumer to buy.

Nevertheless, Freedman firmly believes that RCA's disc players will reach a mass market. The lower price and the availability of much more programming are the important ingredients that he feels will make for success for RCA. Rich's has never been able to obtain more than 100 or so disc titles, out of the more than 200 announced by MCA, and the lack of a steady flow of new titles has been a serious handicap in selling disc players. Freedman likens it to trying to sell new cars when customers know they can only get enough gas to drive five miles a week.

The tug of war between tape and disc is also seen in the experience of Media Concepts, Inc., a St. Petersburg, FL, retailer that has been selling VCRs for several years, but that got its first shipment of Magnavox disc players only in May 1980. Glynda O'Neal, a salesperson with Media Concepts, which has three stores in the area, says the reason consumers would choose a disc player over a VCR comes down to "basically cost." A movie like "Jaws" costs only $24.95 on disc, vs. $65 on tape, and the Magnavox player, at $775, is $300 less than a popular JVC video recorder that the stores sell for $1085.

On the other hand, O'Neal points out, VCR sales haven't suffered at all. Some consumers are able to afford both tapes and discs, but "many people want the ability to record. They keep asking, 'Will they come up with [a disc player] that will record'?" O'Neal concludes, "The fact that it doesn't record definitely puts a crimp in sales" of the disc players.

Another view comes from Mike Weiss of That's Entertainment, a Chicago company that owns two stores selling VCRs, cassettes and discs. He described Chicago consumers as "starved" for discs, since the players were not officially introduced there until September 1980. Any discs that the stores were able to obtain quickly sold out. In the long run, said Weiss, "I can see that discs are going to overtake tape." Lower-priced programming is the key to this trend, along with the higher quality TV picture that some consumers perceive for the optical disc, compared to VCRs.

That's Entertainment has had customers buying 10 discs at a time, and three to five titles are usual—although VCR owners often purchase several prerecorded cassettes too. Because the recording ability of VCRs is important to many consumers, Weiss sees affluent purchasers buying both machines, rather than rejecting tape in favor of disc players. He estimated that 90% of the Chicago consumers who had disc players already owned VCRs.

Discs vs. Cable and Pay TV

Of course, video cassettes and video discs are not the only sources of new video programming available to the consumer. Cable television with its vast channel capacity—up to 80 channels in the newest systems being built—is becoming a formidable supplier of information and entertainment. The pay TV networks like Home Box Office and Show Time that are carried on local cable systems, thanks to satellite transmission, already reached 9.8 million homes by mid-1981. That was four times as many homes as owned a VCR, and about 9,740,000 more homes than owned a disc player.

No doubt it is a great convenience to be able to play a cassette or disc whenever you want, but this convenience must be weighed against the cost: $500 to $1000 for the player, and $20 to $70 per tape or disc, vs. the $20 a month subscription fee for cable TV plus the pay TV channel. For the average consumer to acquire on tape or disc as many movies as he can see each month on Home Box Office would mean a significant increase in the family entertainment budget.

Thus, both VCRs and cable television may limit the success of the disc in the home for the most basic of reasons: money.

CONSUMER DISC PROGRAMMING

Because consumer tastes cover such a wide range, any material susceptible to treatment in an audiovisual format is proper material for the video disc. Broadly speaking, this programming falls into three categories:

1. *Existing entertainment or instructional programs.* Any feature film, TV program, documentary or educational film or filmstrip can be converted to a video disc. MCA had its own Universal Pictures film library, which it drew on to offer such titles as "Jaws" and

"Summer of '42." As discussed earlier, RCA and MCA licensed feature films from MGM, Paramount, Warner and others. Educational films were also an obvious item to offer. MCA has offered programs on whales, sharks and dolphins from "The Undersea World of Jacques Cousteau" as well as "The Solar System/The Universe," "What Makes Rain?/Storms," "Math That Counts" and "Archaeological Dating/The Big Dig." Finally, how-to programs, some of which originally appeared on broadcast television, made their appearance. A typical example was "Julia Child: The French Chef—The Omelette Show" in the MCA catalog.

2. *Newly created entertainment or instructional programs without interactive features.* The reason for producing original programs on video cassette or disc is that such programs become economical when the consumer is paying for them, whereas they would not be economical if the advertiser were paying. This type of program is often labeled special interest, and the following is an example of the economics involved.

An hour of broadcast television entertainment may cost $300,000 to produce, and the 12 minutes of commercial time sold to advertisers in that hour might go for $700,000. If the program only reaches a million viewers, the advertisers have paid 70 cents apiece to reach their intended audience, or $700 per thousand. This may be 50 or 60 times what they normally would pay, and explains why programs that reach "only" a million people can never be aired on commercial TV. But suppose 50,000 of those million are willing to buy the program on cassette or disc, and to pay $25 for it. The resulting gross of $1,250,000 will probably be enough to pay the $300,000 in production costs, perhaps $200,000 to replicate the discs, and another $500,000 in marketing and distribution costs, leaving $250,000 to cover the overhead of the business and to provide a profit. This indicates the economics of special interest video programming, but it remains hypothetical because so few original programs have been produced.

Although MCA's first catalog included how-to and educational programs that might be called special interest, retailers stocking the programs said they did not sell well. However, MCA has not given up; it continues to list how-to programs, and RCA has also licensed some for its disc catalog. It is safe to predict that this form of programming for the disc will grow, but slowly.

3. Original programming in non-linear or interactive formats. Disc experts talk about the difference between linear and non-linear programming, a distinction that has nothing to do with higher mathematics and everything to do with the type of material appearing on the screen. A linear program is a traditional one that tells a story or conveys a lesson, one point at a time. A non-linear program permits jumping around from one segment to another, the repetition of certain sections, the juxtaposition of still frames with motion, and the "branching" to different parts of the program depending on the viewer's interest or aptitude.

As noted in other chapters, the optical video disc lends itself well to this non-linear kind of programming. Researchers at MIT, University of Nebraska, Brigham Young University and elsewhere who have created experimental training or instructional programs are ecstatic about the prospects for education when optical disc players become a staple in schools and homes. It must always be kept in mind, however, that the cost of creating a program has to be recovered through sales of that program in the market, or else the program will not be created in the first place. Thus, researchers working with federal grants or foundation money are not always the best judges of the commercial feasibility of their creations.

One thing that can be safely said is that producers will not rush to pour large sums of money into highly innovative—and speculative— programs for a consumer market that at present scarcely exists. They will first want to establish that consumers are ready and willing to purchase existing entertainment fare on disc, before risking money on original creations. It seems ironic, but inevitable, that the very type of programming that would help make video discs more attractive to consumers must wait until consumers have bought disc players for other reasons. Only then will producers come forward to look at this promising, but largely unexplored, entertainment medium.

CONSUMER DISC MARKET IN 1981

Matsushita and JVC reached agreements with General Electric in the U.S. and Thorn EMI Ltd. in Britain to manufacture the VHD/AHD system. In November 1980, the four companies formed three separate joint ventures: VHD Electronics, Inc., to manufacture the players in the U.S.; VHD Programs, Inc., to license and distribute programming; and VHD Disc Manufacturing Co., to custom press discs for movie studios in the VHD format.

VHD opened a replicating plant in Irvine, CA and planned a second plant in the Midwest to give the VHD format a combined production capability of 20 million discs a year in 1982.

The consortium's timetable called for the VHD players to be marketed first in Japan in October 1981, where the format has already established dominance over the competing systems. Every major Japanese company, with the exceptions of Sony and Hitachi, is expected to bring out a disc player in the VHD format. The players will be launched with an opening library of 100 titles in January 1982 in the U.S. and five months later in the U.K.

As of mid-1981, programming deals had been signed with United Artists and MCA. On March 22, 1981 RCA launched its CED player in about 7000 retail outlets. Though backed by a multimillion dollar advertising campaign the actual dealer support amounted to little more than a series of teaser ads prior to the launch and an ad with dealer listings in newspapers around the country. This effort did not build the kind of instant in-store excitement local dealers expected.

A typical retail reaction was that of the Caldor discount chain in the New York-Connecticut area. Excited by the players' potential, Caldor displayed the players prominently in their stores, but sold few or no players. Within a month after the launch, the chain cut its price from $499.95 to $433. By late June Caldor was selling the RCA player for $388. Small and medium-sized television and appliance dealers had moderate success; video specialty stores seemed less successful and many used traffic drawn in by RCA disc advertising to promote the sales of VCRs instead.

Discs, though more abundant than for the laser optical launch, were still in short supply. Obviously disappointed by the market reaction, RCA began to focus more efforts on retail level promotions. Within three months, RCA had sold 28,000 players and about 220,000 video discs, leaving considerable doubt as to whether it would reach its goal of selling 200,000 RCA brand players and 2 million video discs by year-end.

EUROPEAN DISC PLANS—1981

N.V. Philips originally planned to introduce the laser optical video disc player in the U.K. and Europe in summer 1981. However, this date was subsequently pushed back to autumn 1981 and even further delays seemed possible. The brand name LaserVision was chosen for the system, and the suggested retail price was "under £500." Disc production for the initial catalog of 120 titles was to be in the new Philips plant, built at a cost of £10 million, in Blackburn, England. Instead of acquiring licenses for programs, Philips has been arranging to custom press discs for producers at a charge of £3.20 per disc, in quantity. Magnetic Video, Paramount Pictures, MCA Universal and Rank Video all said they would release discs in the LaserVision format. Philips will distribute the discs to its hardware outlets in Britain, and the discs should retail for an average of £15.

Thorn-EMI planned to replicate VHD video discs in 1982 at a factory in Swindon, England and at its subsidiary EMI-Electrola plant in Cologne, Germany. The factories will have the capacity to produce 3 million discs annually. The targeted launch for the VHD player in the U.K. is June 1982.

As of mid-1981, RCA had no target date for its player launch in the U.K. or Europe. However, it has been actively lining up potential hardware and software partners for eventual European distribution. On the software side, it had reached agreements with Beta Taurus of Germany and Gaumond of France. In England, General Electric Corp. (GEC) will market the CED players, but RCA had not negotiated details of its manufacture.

The prospects for the consumer disc in Europe are as uncertain as they are in the U.S. It is true that there is pent-up consumer demand for video programs; had a long-playing disc been available in 1975, when Telefunken introduced its abortive TeD player, such a product might have had rapid acceptance. But, as in the U.S., explosive growth in VCR sales makes success of the disc questionable. VCR sales in Europe were well over 1 million in 1980 (over 400,000 just in the U.K.) and could reach 2 million in 1981. VCRs are already available for the £500 that the Philips LaserVision player will cost.

The official Philips estimate was for "tens of thousands" of player sales in 1981, and annual sales of 700,000—with disc sales at 18 million units—by 1986. Retailers like Bob Piercy of Piercy's Electronics in London, seemed skeptical. "At that price [£500]," he commented, "it's a very expensive toy."

FUTURE OF THE VIDEO DISC IN CONSUMER MARKETS

In 1981 the consumer electronics industry was poised for a grand battle to establish whether consumers really want disc players, and if they do, which system they prefer. As in any such battle, one cannot expect the opposing generals to be objective about their chances. When RCA's chairman predicts that within 10 years video disc players will be in 30% to 50% of U.S. households—between 23 million and 38 million will be sold in that time—this is not a scholarly conclusion, but the opening shot in a vast marketing and promotion campaign. In the same speech to distributors in which he made this forecast, Edgar Griffiths continued, "When we introduce the Video-Disc, we are going to take over first place, and I guarantee you we will never lose first place." What is at work here is not rational analysis but corporate cheerleading.

Much the same sort of cheerleading, from a different quarter, was at work when U.S. Pioneer displayed its consumer player in spring 1980. Its executive vice president, Ken Kai, asserted that a "large majority" of consumers to whom Pioneer demonstrated its player "were convinced that the laser-optical system, with no disc wear and mechanical distortion, was the video disc system for their future."

Companies like RCA, MCA, Philips, Matsushita and Pioneer have hundreds of millions of dollars at stake in the viability of a consumer market. While these investments will influence consumer attitudes toward the technology, they will not determine those attitudes.

In assessing the future of the disc in the consumer market, it may be best to reduce it to a series of "ifs." The technology has an excellent chance of consumer acceptance only: if manufacturers succeed in making reliable, trouble-free players and delivering both players and accompanying discs rapidly (something that was hardly true in the first year of Magnavox's player and MCA's discs); if the players and discs are significantly less expensive than video cassette records and prerecorded tapes, giving consumers a strong reason to buy them; and if the right programming is available in abundance — programming that appeals to a variety of consumer tastes.

One might argue that to put the case this way is to stack the deck against the disc, but surely skepticism is in order. After all, discs are reaching the consumer market 10 years after they were first demonstrated, and at a time when VCRs and pay TV both have a large head start in the American home. VCRs have a similarly large lead in both Europe and Japan. Success for the video disc in the consumer market would not be a miracle, but neither is it a foregone conclusion.

4

Education and Training Applications of Video Disc Technology

by Paul F. Merrill

Educators and training specialists awaited the introduction of the video disc into the marketplace with great anticipation. As noted in Chapter 3, Magnavox, a subsidiary of Philips, began marketing its consumer model video disc player, Magnavision, in December 1978. (The player is shown in Figure 4.1.) The industrial model produced by Universal-Pioneer (see Chapter 2) was delivered in quantity in 1979. An industrial disc player with similar features is also available from Thomson-CSF.

FEATURES APPEALING TO EDUCATORS

The great interest of the education and training communities in the video disc stems from the fact that each individual frame or picture on the optical video disc can be individually located and displayed in normal motion, slow motion or as a still frame. Still frames may be shown for an indefinite period of time without any wear to the disc or player since the discs are read by a low power, helium-neon laser beam.

Also of great interest to educators is the fact that, in addition to standard film or television programs, many other types of audio-visual or textual material may be recorded on the video disc and dis-

Portions of this chapter were adapted from Merrill, Paul F., and Bennion, Junius, "Videodisc Technology in Education: The Current Scene," *NSPI Journal,* 18(9): 18-26 (November 1979).

Figure 4.1   Magnavision consumer player. Courtesy Magnavox Consumer Electronics Co.

played on a television screen: filmstrips, 35mm slides, microfilm, transparencies, printed text from books, newspapers, computer printouts, etc. Two discrete audio channels are available on the video disc. This provides a stereophonic sound capability or allows the audio to be recorded in two different languages which can be played independently. Still frames, motion sequences or aural information may be combined and intermixed in any sequence on a given optical video disc.

The physical contact video disc systems developed by RCA, JVC, Telefunken and others also have appeal to educators as an inexpensive means of playing prerecorded programs. However, physical contact video disc systems that use a capacitance stylus will have limited still frame capability. These discs will wear out very rapidly if used for still frame disc play, since the stylus will be in constant contact with the tracks that contain pictures which are viewed as still frames.[1]

Special features of interest to educators are available on the Universal-Pioneer (DiscoVision Associates) and Thomson-CSF optical players. These industrial players can randomly access any frame on one side of the video disc. The user may type in a given frame number on a hand-held remote control unit and then press the search key. Within a matter of seconds the player will automatically seek out the specified frame and display it on the TV screen. This frame may be displayed as a freeze frame or as the first frame of a normal or slow motion sequence. JVC's VHD system can provide random access when connected to a special unit (see Figure 4.2).

The industrial players may also be connected to an external computer. Such a combination creates the potential for a very powerful, stand-alone, computer-based education system. The video disc/microcomputer combination offers in one system the media capabilities previously available only in very large and expensive multimedia systems. Motion sequences in full color and stereophonic sound can be followed immediately by a series of still frames. Still frames are available at the touch of a button. Questions appearing on the screen can test the student's knowledge of concepts just presented; the computer can then give the student immediate feedback. Depending on the student's response, the computer can set in motion the appropriate next sequence.

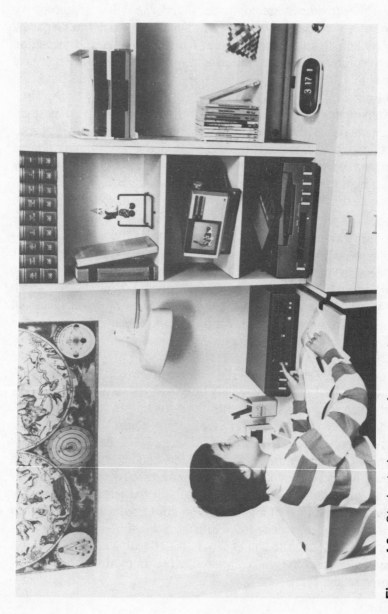

Figure 4.2 Student chooses a frame using a separate random access unit connected to the JVC VHD video disc player. Courtesy Matsushita Electric Corp.

LEVELS OF CAPABILITY

The proliferation of companies producing video discs and players has resulted in several types of players and discs with different operational characteristics, making possible a variety of educational applications. Several authors have attempted to classify different levels of video disc capability or sophistication. (Some players are capable of operating at several different levels.) The following is one possible hierarchy:

1. *Linear play.* The disc is viewed from beginning to end in a linear fashion without stopping or branching. This level is adequate for viewing standard film and video materials and is especially appropriate for entertainment applications. All announced players will be able to operate at this level.

2. *Manual frame access.* The manual controls on the player may be used to locate a particular sequence by skipping forward or back in high speed search mode, to freeze a particular frame, to play in slow motion, to play in reverse, to step forward or reverse one frame at a time, etc. All optical disc systems can be operated at this level. The JVC system will have limited frame access capability.

3. *Automatic stop.* Two different types of stop codes may be encoded directly onto specific frames of the video disc during the mastering process. When the auto-stop code is read, the player automatically switches from normal motion and stops or freezes on a particular frame. When a chapter-stop, which is encoded on 400 consecutive frames, is read during search mode, the disc will automatically switch from search mode to normal play mode. This technique can be used to find the approximate point the user wants to locate, and would be used for each major section of the disc. Although the Magnavision player instruction booklet describes these features, the first set of players marketed by Magnavox did not have the auto-stop feature activated.

4. *Random frame access.* Specific frames may be randomly accessed in less than five seconds. This can be done by entering the frame number on a numerical key pad and pressing the search key.

The industrial/educational optical disc players and the Pioneer VP-1000 consumer player will operate at this level.[2]

5. *Programmed control.* A predetermined series of operational commands and frame numbers may be entered and stored in the 1024 bytes of memory of a microprocessor built into the player. This program will automatically control the operation of the disc by playing in normal or slow motion, freezing on a particular frame for a specific length of time, or branching to an alternate sequence based on a user's response on the numeric key pad. This program of operations and frame numbers may be entered into the memory of the player by using the special programming function keys on the key pad, or by loading the program from the beginning frames of a video disc. This level of operation can be achieved on the DiscoVision industrial/educational players.

6. *Microcomputer control.* The optical industrial/educational players may be connected to a microcomputer. The microcomputer may be programmed to control the operation of the video disc and provide for sophisticated interaction between a user and the disc player.

EDUCATIONAL APPLICATIONS

How can video discs be used to increase the quality and/or decrease the costs of education and training? This question will be answered in this section by describing several applications that correspond to each of the levels of video disc capability described above.

Linear Play

At this level the advantages of using the video disc over other media such as 16mm film or video cassette are fairly limited. If discs are replicated in large quantities, the per copy cost of a video disc may be less than the cost of a film or video cassette. However, in limited quantities the video disc per copy cost would be greater than that for film or video cassette (see section on costs for more detail). The video disc player should be easier to use than a 16mm projector, and video disc programs may be viewed without darkening the room.

The program content for linear play video discs will necessarily be similar to that available on film or cassette.

Manual Frame Access

This level of capability allows the user considerable control over the rate at which video information is presented. This control would be especially valuable in many applications involving the teaching and learning of motor skills. It is difficult for a person to learn a motor skill solely from verbal or textual information; the movements must be demonstrated either by another person or through the use of pictures. In some cases, even a live demonstration is inadequate if the movements are executed too fast for the trainee to see the critical details. For example, the serve in tennis is difficult to demonstrate and learn for this reason.

The manual frame access feature of the video disc will make it possible for trainees to view the tennis serve (or other motor skill) first in normal motion, then in various degrees of slow motion to see the critical attributes. If necessary, trainees may stop on a single frame and view the frozen motion for as long as they wish or they can examine a series of frames one at a time. Research has shown that repetition of a demonstration can improve the learning of motor skills.[3] Such repetitions are easily accomplished with a video disc at this level. Special effects photography would not be required in the production of such programs since the speed of presentation is under the control of the user.

Motor skill applications cut across a broad range of activities from making a pie crust, tying a square knot, welding two pieces of metal and replacing a washer in a water faucet, to installing a sprinkling system, extracting a tooth and performing open heart surgery.

Bennion[1] has suggested that this level of video disc capability could be used for limited interactive instruction. Although branching would be tedious, it would be possible to place sets of still frames—which could be used to present practice questions—throughout the disc. The still frames could be preceded by a motion sequence and a two-to-three-second warning label, just prior to the still frames, instructing the student to stop the motion and view the still frames. The still frames with questions could be followed by appropriate feedback frames. For multiple choice questions, the user could be instructed to step forward a specified number of frames in order to review the appropriate feedback message. The feedback messages could be followed by additional motion sequences and corresponding still frames.

Automatic Stop

The addition of the automatic stop features to the manual access capabilities significantly increases the feasibility of using the video disc for interactive instruction. Auto-stops could be placed on the first frame of each series of still frames used for practice questions and feedback. When the player reads the auto-stop code, it would automatically freeze on the appropriate frame. This feature will eliminate the need for the two-to-three-second warning message described above. Chapter-stop codes would make searching for specific sequences somewhat easier. Thus, it would be feasible to use a menu or limited branching instructional strategy. The automatic stop capabilities are used extensively on the biology discs developed by the World Institute for Computer Assisted Teaching (WICAT, Inc.). (WICAT projects will be described later in this chapter.)

Random Frame Access

The random access capability increases the feasibility of fairly sophisticated branching strategies. However, all branches still must be initiated by the users who would have to be instructed to type in the appropriate frame numbers on the key pad and to press the appropriate search key. An index or menu frame could provide the frame numbers corresponding to each of the available sequences. At the end of each sequence, an auto-stop could be used to show another still frame containing the index to additional sequences.

Random access would also make it possible to implement a sophisticated instructional strategy controlled by the learner and similar to that used with the Time-shared Interactive Computer Controlled Information Television (TICCIT) instruction system, developed by Mitre Corp. in conjunction with Brigham Young University.[4] (See description below.) Through appropriately placed menus or maps, the students could enter frame numbers to select different presentation modes, levels of difficulty or sequences. For example, students could select how many examples to do, whether or not to view help sequences, etc. This type of strategy has been used effectively across many subject areas on the TICCIT system. Also possible is a strategy that instructs students what lesson to study next, based on their response to various diagnostic test items.

This level of capability also gives the classroom instructor greater flexibility in the use of programs designed originally for linear play. A complete film may often contain considerable material that is irrelevant to the topic under study, along with a few short sequences that are relevant. On the disc, such sequences could be readily found and shown in a classroom; the trial-and-error searches and rewindings necessary with a film projector could be avoided. The teacher would merely type in the frame number and access the appropriate sequence within a few seconds. The sequence could be repeated and/or viewed in slow motion or a frame at a time as needed.

Obviously, many of the applications described under previous levels would be easier to implement at this level.

Programmed Control

This level of capability makes it possible for the video disc player to operate automatically under the control of the program entered into the memory of the built-in microprocessor. Thus, the instructional strategies described under the random access level could be implemented without requiring the users to type in frame numbers and press the search key. Users would still be required to enter a number corresponding to their selections, but the program would control which frame to branch to and whether to go into normal play, slow motion or freeze frame mode. For the user, the program removes much of the tedium associated with the random access level. The possibility of errors in entering frame numbers is also reduced. However there is considerable tedium and probability of error involved for the developer using the key pad to enter the program into the memory. These problems can be eliminated by storing the program on the first few frames of the video disc. The program can then be automatically loaded into the memory of the microprocessor when the disc is played.

This level of capability introduces the possibility of an instructor and/or students editing any available video disc to meet their specific requirements. The program stored in memory could be coded to skip over irrelevant sections, completely change the order in which certain sections are shown, modify the speed of presentation, repeat segments, etc.

The instructional strategies implemented at this level are mainly limited by the number of operation commands and frame numbers that can be stored in the 1024 bytes of memory at one time.

Microcomputer Control

This level of capability obviously provides the greatest level of sophistication and produces the most excitement among educators. At this level it is possible to merge the unique capabilities of the book, television and the computer into one integrated medium. This integration has never before been possible, and introduces a whole new set of educational strategies. Applications are limited only by our imagination.

The addition of the microcomputer significantly reduces the memory limitation described under the programmed control section. Not only is the much larger memory of the microcomputer available for operation commands and frame numbers, but peripheral floppy disc drives or audio cassettes can be used to store a large number of student programs and student response data. The full logic capability of the microcomputer is also available to implement very sophisticated branching strategies. The microcomputer can also be programmed to process fairly complicated natural language responses from the user.

One of the most exciting applications that could be implemented at this level is interactive simulations. Simulations allow us to represent a complex real phenomenon or system by a simplified but realistic model. Through interactive computer programs, students would be able to manipulate various aspects of the model and determine the consequences of such manipulations. Manipulation of real phenomena and systems is often dangerous, expensive or extremely time-consuming. Through simulations these factors may be reduced markedly. Speeding up or compressing time simulations allows students to view and manipulate phenomena which may occur too quickly or too slowly in the real world; time compression also allows students to manipulate several variables and see their effects in a very short period, which would be impossible in real time. The sophistication and realism of simulations can be greatly increased through the integration of the visual and sound capabilities of the video disc and the processing and dynamic graphics capabilities of the microcomputer.

Such simulations could be used not only in education and training institutions but also "on the job" to assist both technical workers and professionals. For example, the computer program could lead a technician through a trouble-shooting procedure to find and repair a malfunction in a piece of equipment. The computer could instruct the video disc to display still frame or motion sequences illustrating various possible malfunctions. The technician could then enter data and conclusions into the microcomputer for processing. Once the cause of the malfunction was identified, the microcomputer and video disc could lead the technician through the appropriate procedure for repairing the problem. A similar program could be developed to assist physicians and dentists in the diagnosis and treatment of patients. When a new medical procedure or piece of equipment is introduced, video disc and computer programs could be prepared for on-the-job training.

The section on current projects, later in this chapter, will describe several additional applications that are being developed for this and previous levels.

MARKET FOR EDUCATIONAL VIDEO DISCS

When education is mentioned, we automatically think of the public schools. However, the educational market is much larger than the public schools. Quality educational material is also of interest to private schools, technical schools, colleges and universities, corporations, government, the military, churches, the home and the individual. Each of these units is a potential market for the video disc. We should not place any arbitrary restrictions on where it will be needed and used.

The video disc can be used by a teacher to enhance a classroom presentation, in a learning resource center for individualized instruction, in the library for self-teaching, in the audiovisual center for closed-circuit broadcasts to the classroom, in the laboratory to demonstrate the use of equipment, in the factory for on-the-job training, in the shop as a job aid, in the home as part of a continuing education course or for self-instruction. And the list goes on. Video discs can be purchased by parents, by public school librarians and media center directors, by university professors, by corporate, government and military training departments, by ministers and by continuing education directors.

The video disc can be a valuable aid to teaching and learning across all levels of education from elementary school to graduate training: it's appropriate for the study of any subject that could be enhanced through the use of still or motion visuals.[5] In the sciences, the video disc can demonstrate and simulate natural phenomena; in history, literature and theater, it can help in the analysis of dramatizations; in business, social studies, teacher education and value training, it can demonstrate and simulate human interactions; in physical education and technical training, it can depict motor skills; in mathematics, it can demonstrate physical relationships and show graphs and charts; in music, it can show bowing and fingering techniques. Clearly, there is no lack of applications for innovative, quality educational video discs.

The Nebraska Educational Television Network (NETV) and station KUON-TV commissioned Kalba Bowen Associates to investigate the potential educational markets for video disc programs. Based on its research, Kalba Bowen Associates concluded that colleges would be more likely to purchase video disc players and discs initially than would the public schools, because colleges spend more on A/V materials per institution and per student.[6] However, the public schools constitute a larger market in the long run since their total A/V expenditure is five times that of colleges. The public schools spend more than $1 billion annually for instructional materials (roughly $800 million for textbooks and $200 million for audiovisual materials).[7]

CURRENT PROJECTS

This section presents a brief description of several of the current educational video disc projects being conducted in the United States. It does not include all major projects, but it is representative. Many of the projects described below are using MCA DiscoVision industrial/educational video disc players (Models 700 or P7800). These players are developmental or early prototype models of the Disco-

Vision Associates production model (PR-7820) industrial player.[8]

Brigham Young University (BYU), Provo, UT

The involvement of Brigham Young University in the development of the TICCIT computer-assisted-instruction system led to an early interest in the feasibility of using video discs and microcomputers in combination.[9-11] In January 1979, the McKay Institute at BYU acquired a Universal-Pioneer industrial player and began several video disc projects under the direction of Edward Schneider. Researchers designed and developed a computer link between a microcomputer and the video disc player. Computer programs were developed to control the various functions of the player and to display computer-generated text on a separate screen. In other projects the McKay Institute and several faculty in the BYU Spanish department are developing highly interactive, conversational video discs for language learning.

One disc that has already been mastered incorporates sequences from the classic Mexican film "Macario." The film was selected because of its difficulty level and its linguistic, cultural and artistic interest. The original film was edited down from 90 minutes to fit on a 30-minute disc with approximately 26 minutes of motion and 138 still frames. The still frames are scenes taken from the movie and are used to cover gaps in the story resulting from the edited motion sequences. Forty-five minutes of the movie soundtrack were recorded on a stereo audio cassette tape, with Spanish dialogue on one track and English on the other. The dialogue is synchronized with the appropriate still frames. (The audio cassette is necessary since audio from the disc is muted in still frame mode.) Motion sequences on the disc are also accompanied by synchronized Spanish and English soundtracks on the disc.

The video disc player, audio cassette deck and an interrupt button are all connected to a microcomputer from Billings Computer Corp. (Provo, UT). While watching a motion or still frame sequence, a student may press the interrupt button to stop the movie. When an interrupt occurs, the microcomputer displays several options on the separate display screen. The student may elect to repeat a given sequence with either Spanish or English dialogue, to obtain information on the aesthetic aspects of the production or on the cultural or historical background of the sequence, to go to a glossary of vocabulary items, to obtain a transcription of the Spanish dialogue, to respond to self-test questions related to the sequence, etc. A total of 23 unique choices can be presented to the student, depending on where in the movie the interrupt button is pressed. If a student repeats a still frame sequence, the computer automatically instructs the cassette player and video disc to return to the appropriate location and play the appropriate audio track synchronized with the still frames. (See Figure 4.3.)

Nebraska Educational Television Network, Lincoln, NE

With funds from the Corporation for Public Broadcasting (CPB) and the Bureau of Education for the Handicapped (BEH), the Nebraska Educational Television Network and other cooperating agencies (University of Mid-America, Great Plains National Instructional Television Library and the Barkley Center, all in Lincoln, NE) began a multiphase project headed by Rodney Daynes to develop video discs for instruction. During the first phase a video disc was produced containing closed captions,* using techniques developed by the Corporation for Public Broadcasting.[12]

An elementary school disc on beginning tumbling was developed to play on an industrial/educational player, controlled by the classroom teacher. A secondary school video disc has been developed for Spanish instruction. Students are given access to the industrial/educational video disc player for individualized, self-paced drill and practice activities. A college-level disc has been developed for individualized instruction in educational psychology, by combining an MCA DiscoVision player with a microcomputer such as the Radio Shack TRS 80. The University of Mid-America has produced a fourth video disc, designed to teach the metric system, which will be played on a consumer player.[13]

In addition to the four discs described above, other programs will be developed for those with hearing problems by the Barkley Center/Media Development Project for the Hearing Impaired. These discs will be used to further investigate the potential of video disc technol-

---

*This term refers to the fact that only viewers with a special decoder see the captions, which are intended for viewers with hearing problems.

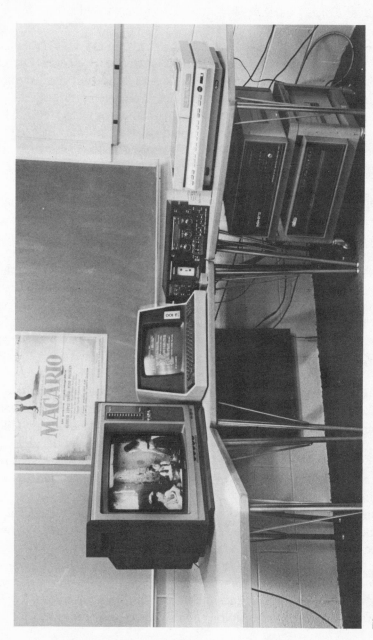

Figure 4.3 Video disc/microcomputer hardware configuration for the Brigham Young University "Marcario" video disc. Courtesy Brigham Young University.

ogy for teaching the deaf. An evaluation of these discs will occur at the Gallaudet College for the Deaf in Washington DC.

Hazeltine Corp., McLean, VA

The Hazeltine Corp. has the marketing rights for the TICCIT computer-based instructional system initially developed by Mitre Corp. and Brigham Young University.[14] At present, students who use the TICCIT system may request that a video sequence be displayed on their TICCIT terminal. This request is fed to an operator terminal. The operator then places a video tape on one of several video tape players that are used with the TICCIT system. Since the present system is operator-controlled, and random access is not possible, video programs are mainly used for introductory segments.

John Volk[15] reports that Hazeltine intends to integrate industrial video disc players with the TICCIT system to replace the present video tape players. The players would make random access possible, significantly reducing the time between a request for a video program and its viewing, and would allow for much higher quality still frame images. These increased capabilities would make it more feasible to use motion sequences and video still frames as examples during a program in addition to introductory segments.

WICAT Inc., Orem, UT

WICAT Inc. has been involved in several video disc projects over the last few years, funded by various private and governmental organizations. WICAT's first major project, funded by McGraw-Hill in May 1977, resulted in the development and mastering of a video disc in biology entitled "The Development of Living Things." This is generally considered the first disc program developed for individualized interactive instruction.[16]

The McGraw-Hill video disc developed by WICAT is especially significant because it was designed to be used on a consumer model player. The disc would take approximately 30 minutes of running time if it were to be played continuously. Two and one-half minutes are used for an introductory motion sequence, 26½ minutes for sound motion picture sequences selected from four different CRM/ McGraw-Hill Films, Inc. biology films, 40 seconds for start-up instructions and 20 seconds for 600 still frames. The 600 still frames are interspersed throughout the disc between several motion and sound sequences.

Special stop codes were placed on the disc during mastering at the beginning of each still frame sequence. When one of these stop codes is read, the player automatically switches from normal motion to still frame mode. The still frames are used for text, diagrams, illustrative photos, photomicrographs, etc.[17] They are also used to present questions to the student, to provide correct answers, to summarize information and to orient the students as to their location on the disc. The orientation information is provided through table of contents frames and special color codes and symbols.

In April 1979, WICAT received a two-year, $400,000 grant from the National Science Foundation to design, build, demonstrate and evaluate an intelligent video disc system. Under this grant, WICAT will evaluate the McGraw-Hill biology disc with different levels and types of students. It will also design a special prototype controller to combine the Magnavox consumer player with a Texas Instruments 9900 microprocessor. This effort will entail some engineering modifications to the consumer player. This microcomputer/video disc system will use two display screens: a color TV monitor for video disc displays and a black and white CRT for computer-generated text and graphics. However, WICAT also plans to develop the capability of overlaying computer graphics on the video disc display for certain applications.

After the microcomputer/video disc system becomes operational, WICAT will also develop and evaluate the effects of computer enhancements to the McGraw-Hill biology video disc. The computer enhancements will include simulations; advisory information in the event that the learner doesn't know how to proceed or is using an unproductive strategy; constructed response and answer processing; more sophisticated orientation procedures than the color codes and symbols discussed above; and student status feedback information, whereby the computer will keep track of how well or poorly the student is doing in the lesson, whether he has solved problems or answered questions correctly, etc.[18]

Other WICAT video disc projects include a study of video disc authoring procedures for the Navy, equipment and production procedures for the National Institute of Education and a cost analysis comparing video discs to video tape for the Army Communicative Technology Office.[19]

Utah State University (USU), Logan, UT

In fall 1978 Utah State University purchased an MCA DiscoVision industrial video disc player and established the Videodisc Innovations Project (VIP) with R. Kent Wood serving as director. A special series of articles describing various VIP activities was published in the May 1979 issue of *Educational and Industrial Television*. VIP also conducted the first National Videodisc/Microcomputer Institute (NVMI), which was held at USU on June 11-15, 1979. The Institute was funded by the U.S. Office of Education.

One of the first VIP projects was to develop an experimental instructional video disc. The project directors decided to incorporate a variety of subject areas and instructional strategies, as well as program formats (e.g., film, tape, slides) onto the disc. This disc could then be used by the project directors to investigate many issues, such as which format would be the best for transferring material to the disc. The formats used included 2-inch video tape, 16mm film, 3/4-inch video cassette, 35mm slides and super 8mm film transferred to 3/4-inch cassette. Subjects included nutritional quality, welding, family life, production of graphics, care of sheep, special education, physics and use of the library.[20]

This experimental disc has subsequently been used in several ongoing investigations. One study linked an IMSAI 8080 microcomputer to the video disc player and connected a touch panel to the TV monitor. The purpose of the instructional program was to teach retarded children to match objects of the same color, size or shape. The computer program was written in the PILOT language with special commands to control the video disc and to detect touch panel responses.

Another study focused on the application of a microcomputer/video disc system for teaching how to use the library card catalog.[21] For this study the video disc player was connected to an Apple II microcomputer. The video disc segment included 90 separate still frames showing interior views of the library and a large number of different types of catalog cards. Future efforts by USU in this area include an investigation into the feasibility of simultaneous display of computer-generated information and video disc images on the same TV screen, and the storage and retrieval of digital information on the video disc.

Massachusetts Institute of Technology (MIT), Cambridge, MA

The Architecture Machine Group of MIT has been experimenting with the development of an information management system that takes advantage of the user's sense of space in the organizing and retrieving of data. This work is directed by Nicholas Negroponte and funded by the Cybernetics Technology Office of the U.S. Defense Advanced Research Projects Agency (DARPA).

The setting of the experimental Spatial Data Management System (SDMS) is a multimedia room about the size of an average personal office. The room contains a wall-sized display screen, a specially instrumented chair, two small touch-sensitive TV monitors and eight loudspeakers providing octaphonic sound. The instrumented chair contains pressure-sensitive "joy sticks" and 2-inch square touch sensitive pads on each arm. An adjacent room contains a TV projector which projects onto the back of the large screen. The system is connected to both computer and video disc equipment. Through this multimedia room the MIT researchers have attempted to create a setting ". . . wherein the user is directly engaged with data . . . inhabiting a spatially definite 'virtual' world that can be interactively explored and navigated."[22]

The spatial world of this information system is called "Dataland." An aerial or world view of Dataland is continuously visible on one of the small monitors, while the big screen displays vastly enlarged sections with far greater detail. The user can navigate around Dataland through the use of the joy sticks on the chair and a touch panel on the small monitor. A small highlighted rectangular area on the monitor indicates the position of the enlarged view seen on the big screen. By manipulating the joy sticks, the user can zoom in on various sections of the picture. The distant views are digital television images which come directly from computer memory or are generated by a computer program, while the close-ups show high-resolution images from a video disc as played on the MCA DiscoVision industrial player.

For example, one data item on the world view monitor might be a small satellite photo of the New England coast, while through the use of the joy sticks the user may zoom in on a selected area such as Boston. As the user gets closer, the digitized satellite image is replaced by a high-resolution map from the video disc in perfect registration, and then, by slide views of a part of the city from different vantage points, selected at random by the user. Many additional examples are described and illustrated by color photographs in Bolt's exciting report.[22]

The MIT group has also developed a bicycle repair course on video disc in which computer-controlled random access capabilities allow the user to examine the repair process at several different levels of detail. The level of detail to be viewed may be selected by the user.[23]

Other Projects

Among other significant projects involving applications of the video disc in education,[24] the Lister Hill Center for Biomedical Communications of the National Library of Medicine, under the direction of Charles Goldstein, is developing an intelligent device controller (IDC) that will allow a computer to control various devices, including the video disc. The Library is also investigating the use of the optical video disc for the mass storage and retrieval of digital information, in order to mix digital, audio and video information on the same disc without difficulty.

Control Data Corp. (CDC) has developed several demonstration programs to show that the video disc could be controlled by a PLATO computer terminal. CDC has developed both a dual display system and a single display system. The single screen display makes it possible to superimpose a reduced resolution form of PLATO graphics over a running video disc program on the same PLATO color terminal.

Video disc players are also connected to PLATO terminals in a project at the University of Nebraska Dental School directed by Gary Jones and funded by the National Library of Medicine. In this project the video disc shows patients in various dental care situations.

The physics department of the University of Utah received a companion grant to that received by WICAT from the National Science Foundation (NSF). This project is an extension of ongoing NSF funding for the development of a computer-assisted instructional system combining video cassette players and microcomputers in an interactive system. Courseware for complete courses in physics and electronics will be transferred from video cassette to video disc.

DESIGN AND DEVELOPMENT OF PROGRAMS

This section will address several of the issues related to the development of interactive educational materials for computer-controlled (i.e., higher level) video disc applications. Development for lower levels would only use some of the procedures outlined for higher levels. For example, non-interactive linear programs (for the lowest level video disc capability) could be developed using the same procedures currently used for films and video tapes.

Bennion[1,25] has suggested that the development of interactive materials for computer-controlled video discs will require integration of techniques currently used in developing motion pictures, slides and computer programs. He recommends that three design forms used in creating films, slides and branching computer programs be combined for designing video disc materials: storyboards, grid frames and branching networks or flow charts. The scripting of motion sequences and slide tape programs is generally represented on storyboards where visual frames or sequences are drawn on the left side of the page and the accompanying dialogue is written on the right side. Slides that are composed of text and graphics are often laid out on grid forms using message design principles to insure readability. The branching logic of computer programs is often represented by flow charts.

Storyboards may be used in scripting the visual and audio sequences of video disc programs, while still frames—either from the disc or as generated by a microcomputer—may be designed on grid forms. Grid forms will be especially helpful for designing computer-generated still frames since the number of characters that may be displayed on the screen at one time is severely restricted (usually 35 to 80 characters per line and 16 to 24 lines per page). An interactive video disc program will also require sophisticated branching logic. Flow charts or branching network diagrams may be used to show the many alternate paths through the various video disc components such as motion sequences, still frames and computer-generated displays. A symbolic coding scheme would need to be used to label these various components. These labels could then be used on the flow diagrams to show the sequential relationships between components.

A special form (see Figure 4.4) was developed for scripting a computer-controlled video disc program at BYU, which incorporates aspects of all three of the forms described above. This special form has a section to describe a short motion sequence, a section for the audio accompanying the motion sequence, a section showing the layout of a still frame which follows the motion sequence, and so on. These forms can be used by authors and reviewers to analyze the content and logic of the program prior to its actual production.[1]

Of course, preparing a script should come only after completing standard instructional design procedures, such as the identification of goals, assessment of needs, content analysis and media selection.[26]

After script forms have been completed, the separate components of the video disc program may be selected from existing materials or produced afresh. The video disc motion sequences and still frames will be assembled onto one medium using either film or video editing equipment. The edited film or video tape is then sent to a mastering facility for the preparation of a video disc master and the replication of video disc copies.

The computer program that will control the video disc can be completed after the replicated discs are produced. Having the video disc copy will insure that all branching addresses are actually consistent with the appropriate frame numbers on the replicated disc. The computer program may also be modified to change the order in which the disc sequences are presented if pilot testing of the integrated video disc/microcomputer program reveals that this would be advantageous. If the control program is to be stored on the first few frames of the video disc, then great care must be taken to make sure the control program frame references correspond to the actual disc frame numbers before mastering takes place.

The art of instructional design has not yet reached the point where an effective instructional program can be guaranteed by following a set formula. Thus, almost all instructional design models stress the importance of formative evaluation, involving pilot testing a prototype version with actual students. Data from these pilot tests are then used to revise the program to make it more effective. Unfortunately, formative evaluation of interactive video disc programs is significantly more complex than that of many other media programs. Since the mastering process is relatively expensive and discs cannot be modified

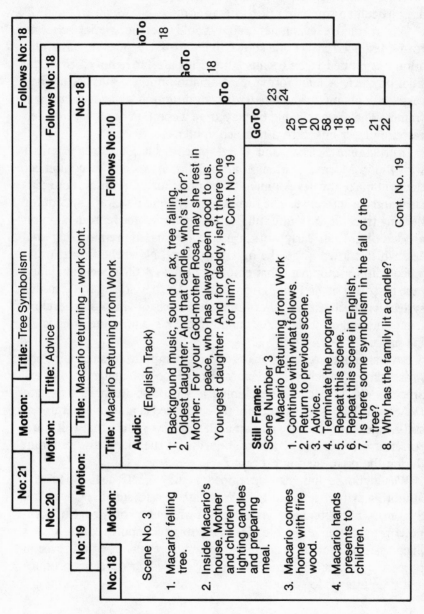

No: 21   **Motion:**   **Title:** Tree Symbolism   **Follows No: 18**

No: 20   **Motion:**   **Title:** Advice   **Follows No: 18**

No: 19   **Motion:**   **Title:** Macario returning – work cont.   **No: 18**

No: 18   **Motion:**

Title: Macario Returning from Work   **Follows No: 10**

Audio:

(English Track)

1. Background music, sound of ax, tree falling.
2. Oldest daughter: And that candle, who's it for?
   Mother: For your God-mother Rosa, may she rest in
   peace, who has always been good to us.
   Youngest daughter: And for daddy, isn't there one
   for him?   Cont. No. 19

Still Frame:
Scene Number 3
   Macario Returning from Work   **GoTo**

1. Continue with what follows.   25
2. Return to previous scene.   10
3. Advice.   20
4. Terminate the program.   58
5. Repeat this scene.   18
6. Repeat this scene in English.   18
7. Is there some symbolism in the fall of the
   tree?   21
8. Why has the family lit a candle?   22
                                        Cont. No. 19

Scene No. 3

1. Macario felling
   tree.

2. Inside Macario's
   house. Mother
   and children
   lighting candles
   and preparing
   meal.

3. Macario comes
   home with fire
   wood.

4. Macario hands
   presents to
   children.

GoTo

23
24

GoTo

18

GoTo

18

Figure 4.4   Scripting form for computer-controlled video disc. Sample script is taken from Brigham Young University's "Marcario" video disc. Courtesy Brigham Young University.

after mastering, it is important that formative evaluation take place prior to mastering. However, the unique capabilities of the video disc are difficult to simulate prior to mastering.

One formative evaluation phase could use the special scripting forms described earlier and shown in Figure 4.4. However, this evaluation has limited value because the forms are far removed from the final product. A later formative evaluation phase could use a video cassette containing the edited video disc motion sequences and still frames. The cassette could be played on a computer-controlled video cassette player linked to a microcomputer.

Instructional Science and Development, Inc., under a contract with Philip Morris, is attempting to define new types of equipment that will make the development of graphics for video discs more efficient and cost effective. The firm is also trying to develop a system for keeping track of the many still frames developed for reproduction on a video disc. In its early work it used a film-based premastering process which turned out to be quite expensive. Now, the firm is using video editing equipment that employs SMPTE* time codes on video tape to isolate single frames from a tape. The developers feel that this system will decrease the labor-intensive aspects of video disc development and make formative evaluation possible during the premastering process.[27]

Several other organizations are also attempting to streamline the process for producing and premastering video disc materials. For example, WICAT, Inc. is developing a Videodisc Authoring System (VAS) which consists of a computer-controlled character generator, a still-frame storage device and an author review system that will allow for electronic review of all components of the video disc program during the premastering process.

Westinghouse has also developed a simple, efficient video disc authoring system called WICS (Westinghouse Interactive Compiling System). This system allows developers without a knowledge of computer programming to write video disc/microcomputer courseware.[28] Such systems should facilitate the formative evaluation and revision process, reducing the time and costs required to develop, produce and premaster video discs.

---

*Society of Motion Picture and Television Engineers.

COST FACTORS

The integration of video disc and microcomputer technology could have a significant impact on education, but what are the costs? The video disc cost picture offers both good and bad news. The good news is that disc replication costs are low, and video disc player costs are competitive. The bad news is that production and mastering costs are relatively high. This section will briefly address each of these cost factors.

Development and Production Costs

Total costs for developing and producing educational video discs are very difficult to obtain. Many organizations do not track and report all of their costs. In some cases production costs are reported but developmental costs are ignored. When data on specific costs are available, we find that the costs vary widely because of different development and production procedures and different instructional strategies. Very few educational discs have been produced to date, and the cost associated with their production includes a lot of trial-and-error learning costs. A detailed description of some of the cost factors involved in video disc production may be found in Mendenhall.[29] The following is a brief summary of that report.

The costs for producing motion sequences for video disc are basically the same as for standard film and video tape programs. The production of good quality original motion footage can run from $1000 to $4000 per minute. These figures may not include front-end instructional design costs. Motion footage costs can be reduced by using existing footage. For example, Encyclopaedia Britannica Films sells film clips from its "Encyclovideo" file for $216 per minute or for a discount price of $75 per minute if the film clips are to be used in local programs only.

The cost for producing still frames also varies widely. Photographs are relatively inexpensive. However, it is estimated that still frames which require original art work and revisions cost approximately $100 per frame. This cost can be reduced significantly through the use of a character/graphics generator designed to produce slides. In general, still frames are more expensive than motion frames. However, Bennion and Schneider[10] have shown that if more than 20% of the disc is composed of still frames, the cost of instruction per hour is basically constant since the length of instruction increases with the addition of still frames.

After the component motion sequences and still frames are produced, they must be edited together in the appropriate sequence. This is the premastering process, and it requires the use of sophisticated editing equipment as described in the previous section. The process results in additional cost.

Mendenhall[29] presents two hypothetical examples of the component and total costs associated with the production of video discs. In his first example the minimum production cost is $23,080 for a 30-minute disc with 600 still frames, using existing footage for motion sequences and a character/graphics generator to produce still frames. The second example has a total cost of $81,100 and includes original motion footage and still frames which are typeset or contain original art. These figures do not include the front-end instructional design costs, which Mendenhall estimates would run at least another $23,000.

If the video disc is to be controlled by a microcomputer, the production costs would also include the computer programming costs. The standard cost for producing computer-assisted courseware on the PLATO computer system ranges from $50 to $200 per hour of instruction depending upon the skill of the author/programmer and the nature of the computer program.

As can be seen from the above discussion, it is extremely difficult to provide a simple estimate for development and production costs. The number of variable factors is too great. Kalba[30] cites several different authors who estimate development and production costs ranging from $5000 to $700,000 for a single program. The low estimate involves assembling materials from existing sources with minimal front-end costs, while the latter includes expensive original broadcast quality motion sequences.

Mastering and Replication Costs

An optical video disc master is produced by exposing the photoresist material on a revolving glass plate with a laser beam. The laser beam is controlled and modulated by a master video tape that contains all the edited motion sequences and still frames. The master video disc is used to produce several nickel stamper sub-masters.[29] This mastering process requires fairly sophisticated and expensive equipment.

The nickel stampers are used to press or form replicated copies into mylar or plastic, using a process similar to that used for producing audio records.[10] Theoretically, this process should result in very low cost duplication. Early promotional literature from MCA DiscoVision projected a manufacturing cost of $.60 per disc.

However, the mastering and replication process has experienced considerable technical problems, resulting in higher prices than were originally expected. The latest quotations from DiscoVision Associates show that the price for mastering and replicating one copy of a two-sided video disc with 30 minutes per side is $3500. The price for mastering and replicating 3000 copies of a two-sided disc is $33,000 or $11 per disc (see Table 4.1). The price for a single-sided disc would be a little more than half the two-sided price. Supposedly, the per copy cost would be reduced as the number of replications increased. Table 4.2 shows the estimated comparative per copy costs of video discs and other motion visual media.

Equipment Costs

Original price projections for the consumer model optical video disc player and the industrial/educational player were $500 and $1000, respectively. However, the present price quotations are considerably higher. Current estimated prices[31,32] for several video disc players are listed in Table 4.3 along with current comparison costs for other audiovisual projection equipment.

Cost Summary

The estimated costs of educational applications of the video disc are potentially volatile because of the infancy of the industry. Competition and mass production could bring the prices down; inflation and poor market reception could drive the prices up. There are also additional costs not previously mentioned: packaging, distribution and marketing; equipment maintenance; storage and security; microcomputer hardware and maintenance; administrative and overhead, etc. The financial viability of marketing educational video discs

Table 4.1   DiscoVision Associates Industrial Prices

Product/Service	Price	Description
Player – Model 7820		
Quantity: 1 to 4	$3,000	
5 to 99	2,250	
100 or more	2,000	
Carrying case	300	
Demo disc	$7.50	
Replication		
Prices per side		
Type I CAV 27 min	5.00	No programming linear. (Basic disc)
Type II	6.00	One program dump. Up to 30 additional frames.
Type III	7.20	One program dump. Over 30 additional frames.
Type IV	8.65	Two program dump. Up to 30 additional frames.
Type V	RFQ*	Two program dump. Over 30 additional frames.
Blank side	2.50	
Minimum order		50 sides replicated per master
Mastering		
Prices per side		
Type I	$1,500	
Type II	1,650	
Type III	1,815	
Type IV	2,000	
No set-up charge		
Maintenance service		
On-site service	$132/yr.	
Depot service	100/yr.	

Pre-mastering service (preparing client media for mastering) per hour charges, plus material, apply to all orders.

Types of discs can be intermixed and a single-sided disc must have a blank side.

*Request for quote.

Uniform freight charge of $46.50 per player.

All prices listed above are subject to change and should be reviewed with a DiscoVision representative before use.

Effective 7/1/80

Table 4.2   Per Copy Costs for One-hour Motion Visual Program*

Medium	Quantity			
	1	10	100	1000
16mm film	$3,591	$624	$242	$187
¾-inch video cassette	65	55	50	47
½-inch video cassette	67	37	27	23
Video disc	3,500	350	40	13

*These costs are based on the assumption that the original program material is in video format.

depends upon the availability of quality software that meets the needs of a particular set of users at a price that does not exceed the perceived benefits gained.[30] Theoretically the cost of quality video disc courseware could be very low if the size of the market is large enough. Five hundred thousand dollars ($500,000) in total costs (including profits) could be recovered by selling 100,000 discs at $5 each (assuming that at this quantity replication costs drop to the level of $1 or $2 per disc). At $5 per disc, each student or home becomes a potential customer of educational video discs.

Table 4.3   Costs of Motion Visual Reproduction Equipment

Equipment	Type	Unit Cost
Magnavision 8000 video disc player	Optical, consumer	$ 775
Pioneer VP-1000 video disc player	Optical, consumer	750
DiscoVision PR-7820 video disc player	Optical, industrial	3000
RCA SelectaVision video disc player	Stylus, consumer	500 (approx)
Bell & Howell 1592 16mm projector	Self loading	773
Elmo GS800 super 8mm projector		849
Fairchild Seventy-07 8mm projector	Rearscreen, cartridge	459
Sony U-Matic VP 2010 video cassette player		1775

CONCLUSION AND SUMMARY

The video disc/microcomputer technology provides great promise for increasing the efficiency and effectiveness of education. However, the realization of that promise is contingent upon many factors. Educators have hoped for years that electronic technology would have a significant impact on education, but that impact has been limited and slow in coming for a variety of reasons. The costs have been too high, the availability of quality courseware has been low, and there has been considerable resistance to change.

The size of the educational market for video disc/microcomputers will be largely determined by two factors: 1) the quantity of good programming available and 2) cost. Both factors are contingent upon the optical video disc's success in the consumer market. A mass market for optical disc players, along with continued advances in technology, will reduce hardware costs and probably result in more, and better quality, programming. However, the mass market for optical video disc players is open to question at the present time. There is considerable competition from consumer video cassette players, which were established in the market first. Cassette players, which offer recording capability and extended play time, are decreasing in price and increasing in capabilities.

In addition to video cassette players, the optical video disc will have strong competition from the capacitive video disc players (using a stylus) such as RCA's SelectaVision, which was introduced at a price of $499 in March 1981. It may be difficult to convince consumers to buy the optical player at the considerably higher price. This factor would limit the size of the educational market for the optical player, unless reported advances—e.g., less expensive optical pick-up components—can bring about lower prices.

In the long run, hardware capabilities and costs will not be the limiting factor to adoption of the video disc. The critical difference will be in the development of creative and innovative courseware. With a mass market, the incentives will be there. However, the task will require significant capital investments and our best creative talent. Only compelling software will establish the video disc in the education and training world.

REFERENCES

1. Bennion, J.L., "Authoring Interactive Videodisc Courseware," in M. DeBloois (ed.), *Creating Courseware For Microcomputer/Videodisc Learning Systems* (Englewood Cliffs, NJ: Educational Technology Publications, in press).

2. *Videoplay Report* (Danbury, CT: C.S. Tepfer Publishing Co., March 31, 1980).

3. Hoban, F. and Van Ormer, E.D., *Instructional Film Research, 1918-1950* (Fort Washington, NY: U.S. Naval Special Devices Center, 1950).

4. Merrill, M.D., Schneider, E.W. and Fletcher, K.A., *TICCIT* (Englewood Cliffs, NJ: Educational Technology Publications, 1980).

5. Merrill, P.F. and Bunderson, C.V., "Preliminary Guidelines for Employing Graphics in Instruction." Paper presented at the Annual Conference of the National Society for Performance and Instruction, Washington, DC, April 1979.

6. Videodisc Design/Production Group, "A Summary of Research on Potential Educational Markets for Videodisc Programming." Project Paper 1 (Lincoln, NE: KUON-TV/University of Nebraska—Lincoln, November 1979).

7. Quirk, D., *The El-Hi Market, 1980-1985* (White Plains, NY: Knowledge Industry Publications, Inc., 1979).

8. *Videoplay Report,* June 25, 1979.

9. Bennion, J.L., "Possible Applications of Optical Videodiscs to Individualized Instruction." Technical Report #10 (Provo, UT: Institute for Computer Uses in Education, Brigham Young University, February 1974).

10. Bennion, J.L. and Schneider, E.W., "Interactive Videodisc Systems for Education," *Journal of the Society of Motion Picture and Television Engineers (SMPTE),* 1975, 84(12): 949-953.

11. Schneider, E.W., "Videodiscs, or the Individualization of Instructional Television," *Educational Technology,* May 1976: 53-58.

12. Daynes, R., letter to author, June 1979.

13. Videodisc Design/Production Group, "Videodiscs Unveiled at an NAEB Convention," *Videodisc Design/Production Group News* (Lincoln, NE: KUON-TV/University of Nebraska—Lincoln, February 1980).

14. *An Overview of the TICCIT Program* (McLean, VA: Mitre Corporation, 1974).

15. Volk, J., telephone conversation with author, June 1979.

16. *Videoplay Report,* June 11, 1979.

17. *Videoplay Report,* July 24, 1978.

18. Bunderson, C.V., personal interview, June 1979.

19. McRae, J.L., "Economic Analysis of Videodisc Technology in the Delivery of Exported Training Packages," *Proceedings of the Conference on Interactive Videodisc and Media Storage Technology in Education and Training* (Warrenton, VA: Society for Applied Learning Technology, 1979).

20. Willis, B.D., "Formats for the Videodisc—What are the Options?" *Educational and Industrial Television,* May 1979: 36-38.

21. Wooley, R.D., "A Videodisc/Portable Computer System for Information Storage," *Educational and Industrial Television,* May 1979: 38-40.

22. Bolt, R.A., *Spatial Data Management* (Cambridge, MA: Massachusetts Institute of Technology, Architecture Machine Group, 1979).

23. Negroponte, N., "Introduction To Dataland." Paper presented at the Annual Convention of the Association for Educational Communications and Technology, Denver, April 1980.

24. Merrill, P.F. and Bennion, J.L., "Videodisc Technology In Education: The Current Scene," *NSPI Journal,* November 1979: 18-26.

25. Bennion, J.L., *Authoring Procedures For Active Videodisc Instructional Systems* (Provo, UT: Institute for Computer Uses in Education, Brigham Young University, March 1976).

26. Gagne, R.M. and Briggs, L.J., *Principles of Instructional Design* (2nd ed.) (New York: Holt, Rinehart, and Winston, 1979).

27. Kribs, D., telephone conversation with author, June 1979.

28. Hoffman, B., "TTO Develops New Interactive Training System," *Westinghouse Defense News*, June 1979: 16.

29. Mendenhall, R.W., "Authoring System Alternatives For Educational Uses of Videodisc Systems." Report submitted to the National Institute of Education (Orem, UT: WICAT, Inc., April 1979).

30. Kalba, K.K., "Reproduction, Distribution, and Utilization Costs For Programmable Videodisc Technology." Phase I Task Report submitted to the Nebraska Educational Television Network (Cambridge, MA: Kalba Bowen Associates, November 1978).

31. *Videoplay Report*, April 14, 1980.

32. *Videoplay Report*, May 26, 1980.

5

Creating Educational Programming for the Optical Video Disc

by Kenneth S. Christie

The exploration of educational applications for video disc technology is in its infancy. The hardware itself has been available for a short time but it is only since the late 1970s that any attempts have been made to design and produce material specifically for the optical video disc. Most of the discs available today are a direct transfer of existing audiovisual fare. There are, however, a number of experimental video disc projects around the country, whose central aim is to develop and implement educational software that takes advantage of the disc's unique capabilities. This chapter will concentrate on specific scripting and production techniques used by the Nebraska Videodisc Design/Production Group.

As discussed in earlier chapters, the optical video disc offers such features as random access, variable motion and programmable capabilities. Other A/V formats, such as film, slides and audio tape, can be combined onto the disc, and the operation of the player can be controlled by the user. It is the combination of these capabilities and the disc's unique characteristics that holds out the promise of programs that educate better than existing A/V formats. Such programs could be applicable to students at all levels—primary, secondary, college and continuing education.

When one considers the microprocessor programming available on some video disc players and the ability to connect the video disc player to an auxiliary microcomputer, there is the possibility for truly interactive instruction using the video disc. Such instruction could revolutionize the world of instructional media.

A PILOT PROJECT

Efforts to design and produce educational video disc programs have thus far been limited. However, a significant framework for educational video disc development has been established by approximately a dozen projects nationwide, including those at MIT, WICAT, University of Utah, etc. (See Chapter 4.)

One of the most extensive efforts, in terms of the scope of research and variety of programs produced, is being carried on by the Nebraska Videodisc Design/Production Group in Lincoln, NE. According to Project Director Rod Daynes, the Nebraska Group is a "specially created unit of designers, producers, engineers and video support personnel dedicated to the development and production of innovative programs that use the video disc's unique capabilities."[1] Formed in 1978 under a multiple-year grant from the Corporation for Public Broadcasting to KUON-TV/University of Nebraska-Lincoln Television, the group has two goals: 1) to produce video discs for varied educational and training applications, and 2) to disseminate research on video disc design and production procedures and on potential educational and industrial markets for video disc.[2]

The Nebraska Group is producing pilot programs for all educational levels and was planning to have 21 sides (over 10 hours) of video disc programming completed by the fall of 1980. The subjects include elementary tumbling skills, high school Spanish, college-level educational psychology, metrics, dental surgery, flight simulation, physics and medical education. In cooperation with the Media Development Project for the Hearing Impaired (MDPHI),[3] an activity of the Barkley Memorial Center at the University of Nebraska-Lincoln, the Nebraska Group has assembled many video discs for learners with hearing difficulties, based on instructional design and scripts from the MDPHI. The Nebraska Group is also pioneering development of interactive video disc materials by means of a link between the video disc player and a personal computer. Several programs have been developed on dental surgery for use with the Programmed Logic for Automatic Teaching Operations (PLATO) computer system; programs on behaviorism and flight simulation have been designed for use with a Radio Shack TRS-80 computer. Finally, Nebraska Group engineers Darrell Schweppe and Hermann Siegl have developed the first true video "interlace" that combines computer graphics and video disc display on the screen of a standard TV set.

EDUCATIONAL APPLICATIONS OF THE VIDEO DISC

One way to understand the educational market for video discs is to look at those educational applications that best use the video disc's unique capabilities. Certain video disc player functions suggest specific types of program uses.

For example, dual channel audio suggests the placement of English on one channel and a foreign language on the other channel. This could be especially useful in high school and college level language courses for vocabulary study and conversation development. Dual channel audio also presents the opportunity to have separate program instructions for the students and for the instructor. Each would be on a separate audio track and only the one needed at the time would be accessed.

One might think that the microprocessor programming ability of the video disc suggests programs for higher educational levels only, but there are many cases where complex video disc programming can improve learning for younger students. The video disc can be produced and utilized in such a way that students think they are playing a game when in fact they are receiving instruction. One excellent example of this type of learning has been designed by the Media Development Project for the Hearing Impaired. This program consists of two video discs on thinking skills for upper elementary students with hearing difficulties. One disc is organized around the concept that the student is a detective who must solve a mystery based on the facts presented to him. The other disc allows the student to solve the problem of being lost away from home. The video disc player is connected to an auxiliary microcomputer that generates all the alphanumerics and controls the disc player. The student is asked which path he wishes to pursue in an attempt to solve the problem. Some paths lead to solutions, some do not, but the student always has the option of backtracking and picking another path if he determines the one he is on is not correct.

This learning approach, while entertaining, is designed to teach a basic problem-solving process. Indeed, virtually any functional use of the video disc that improves the educational approach of a particular subject should be examined and evaluated. Little experimentation with the video disc has been completed thus far to demonstrate which approaches actually improve instruction at various educational levels. For example, variable motion can make certain programs more visually enticing and thus motivate students who might otherwise lack enthusiasm. The Nebraska Group examined this concept in field testing its "Basic Tumbling Skills" disc with elementary pupils in Lincoln. There were some students in the physical education classes who never really cared to learn anything about gymnastics, but after they saw each skill precisely demonstrated by peers, using slow motion and still frames, their class participation increased and their own performance improved.

DESIGNING AND PRODUCING VIDEO DISCS

Because many educational and industrial producers are interested in program production information, an understanding of the complete process involved is needed. The following paragraphs detail what it really takes to put a video disc program together.

The Nebraska Group calls the process premastering. It includes all stages of program development and production up to actual video disc mastering: design, development, assembly and evaluation of video disc programs. The premastering process is different from other program production and there are special considerations during each stage of the process.

Design

The first stage is design. It is here that the instructional designers work closely with a client to determine the educational objectives of a program, including the specific lessons to be learned and the educational level at which the disc will be used. This process occurs for any instructional program, but an instructional video disc program can have more than one set of educational objectives and a single disc can be designed for use by different educational levels. This is possible because of the disc's random access and dual language capabilities. Certain segments designed for use by quick or more advanced learners need never be seen by a slower group, and each audio track could contain differing levels of sophistication. The instructional design also includes how much interchangeability the disc will have; i.e., whether it is designed for use with more than one level of disc player.*

Scripting

The design stage also involves the scripting process, which requires great attention to detail. Because a video disc program can contain both motion and still sequences, as well as two audio tracks, and can be designed for interchangeable play on different types of video disc players, a conventional television script style proves ineffective.

Many groups have modified a script style developed by Junius Bennion of Brigham Young University.[4] It assigns a reference number to every element of a program, whether a still frame or motion sequence. Each element is further broken down to include separate information for each audio track. This greatly simplifies the production and editing process and also allows for the inclusion of other programming options such as captioning (closed or open) for hearing-impaired learners.

The other important consideration in scripting is that it must specify the formula to find the best order in which segments are assembled on the disc. This can become critical if the program is used for interactive instruction. The formula establishes which sequences will be accessed the most and in what order. The program segments are then assembled accordingly to reduce screen-blanking (the amount of time the player needs to search out the next needed segment).

Video disc script format

Figure 5.1 is an excerpt from a program entitled "By Yourself," a visual textbook containing a film, a song, two poems, and a vocab-

---

*The Nebraska Group has identified four player performance levels: The Basic Player, any consumer model player (either video tape or video disc) designed for real time program play; Level I, a consumer model video disc player with freeze frame, auto stop, chapter search and dual audio, but with limited random access, limited memory and no processing power; Level II, an educational/industrial model video disc player with the capabilities of a Level I player, plus a built-in programmable memory; Level III, the same educational/industrial model player connected to a computer.

VIDEO   **CAPTIONS**

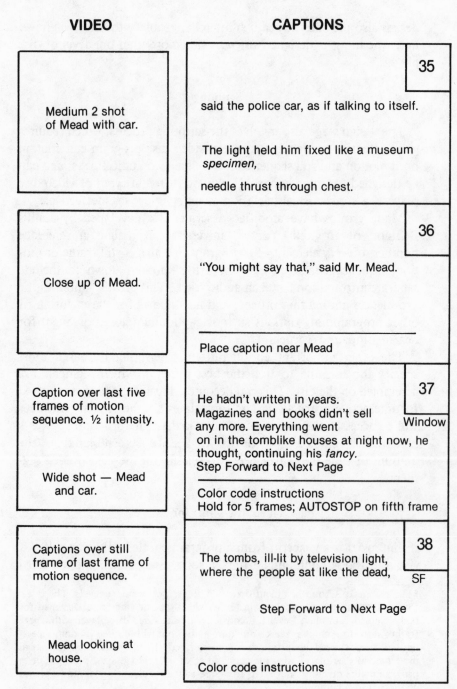

Figure 5.1   Video disc script format.

AUDIO TRACK 1   **AUDIO TRACK 2**

	35
NARRATOR: The light held him fixed like a museum specimen, needle thrust through chest.	
Presence of the man and the car.	Only the presence of the man and the car.

	36
Mead: "You might say that."	
	Only the presence of the man and the car.

Window — No Audio	Window — No Audio 37

	38
Still frame — No Audio	Still frame — No Audio

Figure 5.1   Video disc script format (cont.).

ulary section. The film is based on the Ray Bradbury short story, "The Pedestrian." (This program was developed in cooperation with the Media Development Project for the Hearing Impaired located in the Barkley Memorial Center at the University of Nebraska-Lincoln.) Every sequence is numbered and is designated as either motion or a still frame (SF). Also, each sequence is split into video, graphics and two separate audio tracks. Video and graphics are separated to allow for sequences with captioning over video for hearing-impaired viewers, and/or still frame alphanumerics over a separate video background. Note that each audio track can be used for stereo or multiple sound sources. This program is designed to motivate hearing-impaired students to read, so the motion sequences periodically stop and require the student to read a few pages (still frames) before he or she can continue with the next motion sequence.

Development

The development stage is where production takes place. Studio and remote segments are recorded and the video levels (the level of video indicated on a waveform monitor) for these segments must be checked closely to assure that they do not exceed normal limits. This is critical because a program will go down at least three more generations before it gets to the video disc.* High or peak video levels (a reading of 100% on the waveform monitor) can cause problems when the disc is mastered and may result in rejection of the master tape by the mastering facility. (If saturated color backgrounds are used, readings should be kept well below 100%.)

Film segments can be produced, too. If slow motion use is anticipated in the program, it is recommended that they be shot on film at 30 fps (frames per second) and transferred to tape at that rate. This eliminates the flicker which results from slow motion use of disc segments originating on tape. (Flicker occurs because each frame of video tape has two fields, which are separate images. The video disc player displays both fields when freezing on a frame. If there is motion between the two fields, a flicker results.)

---

*Each separate tape of a program created in the production, editing or duplication process is a generation.

Electronic graphics, with special conventions designed to aid the user, are also generated at this time. Program commands and frame reference numbers can be placed in strategic areas on the display to inform the user of where he is in the program and to tell him what he needs to do to continue. (See Figures 5.2 and 5.3.)

Finally, this stage includes the development of computer programming if the production is designed for computer control (from either the player's internal microprocessor or an external microcomputer). The programming software must be written at this stage so that the appropriate codes can be placed on the master tape before it is sent to be mastered.

Assembly

Assembly of video disc programs is best effected on video tape. It is easier and less expensive to work with than film. And it is essential that the master tape be assembled using a computer-controlled editing system. That is the only way that sequences of single frames of video can be recorded. They could be built up on film and then transferred to tape, but the current transfer method for film to video disc (i.e., the method used by DiscoVision Associates) does not lock down one frame of film to one frame on the video disc. Instead it utilizes a three-to-two pulldown method in which every other frame of film occupies three video fields instead of two. If still frame sequences are built up on film, they must be transferred back to tape at 30 fps, instead of 24. This requires special equipment and extra cost, but will eliminate the problems on the video disc.

Evaluation

Once the program is complete, it is evaluated under test conditions. Each of the original instructional objectives is evaluated and changes are made if needed. It is crucial that this evaluation occurs prior to mastering, because once the disc is mastered, it is permanent; corrections cannot be made. Sometimes this evaluation involves the use of a video disc simulator, a special video tape machine that operates like a video disc player and displays the information in a similar fashion.

This completes the premastering process.

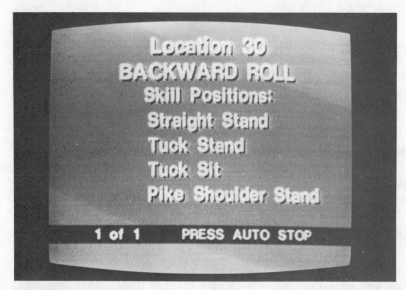

Figure 5.2   Frame showing user where he is on video disc. Note instruction at bottom of screen. Courtesy KUON-TV/University of Nebraska-Lincoln.

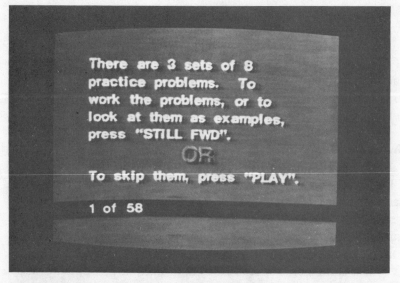

Figure 5.3   Instructional program on video disc telling user what to do next. Courtesy KUON-TV/University of Nebraska-Lincoln.

CHANGES LIKELY

It is important to reiterate the fact that the video disc industry is only beginning to develop. The hardware is constantly being updated and revised. Many new player models will shortly reach the market. Original programming is still scarce. As the hardware progresses and improves, the applications for software will change, too.

Video disc players may become more sophisticated with advanced programming capabilities. There are many possibilities being discussed. A so-called smart video disc player may be developed that could constantly monitor a student's progress for a particular instructional lesson. By a technique known as branching, it could either move the student ahead to more advanced problems if he demonstrated mastery of the basic lesson, or it could take him through remediation sections on the disc if he needed more help on a particular problem. The player's microprocessor could also be equipped to print out a report of the student's progress for the instructor, who would then evaluate the student's progress and determine the appropriate action to take. The printout could even contain information on the rate at which students learn particular lessons—information of value for research on educational strategies and for pinpointing problem areas common to a majority of students.

Perhaps a player with two (or more) heads will be developed. While one head played a particular segment, the other would search out the next needed frames and wait patiently to display them. Although it only takes four seconds or less to search out any position on the disc, a player with two heads could reduce that access time to zero for a completely computer-controlled program. This could speed up individual interactive instruction considerably.

Compressed audio is also a distinct possibility. The idea of placing 10 to 15 seconds of audio on each single frame of a disc is already being experimented with. If compressed audio were placed on every frame of a disc, that could mean more than 200 hours of programming per side! Without pondering the time and energy involved to produce such a disc, consider its possible uses. Language dictionaries that not only spelled a word and presented the definition, but also pronounced the word correctly for the user. Or an art history library with descriptions of each painting and quotes by the artists themselves. Perhaps a complete course in music education. Put such a program under computer control and it could advance automatically to string frames together when one frame of compressed audio was not sufficient to explain a particular idea.

There are even more sophisticated possibilities. A medical group is contemplating the linking of the educational/industrial video disc player to a computer-controlled medical training dummy. A light pen touched to specific points on the dummy would tell the computer to search out pertinent information on the video disc. This application could enable more medical students to obtain hands-on training (and could reduce the need for the number of cadavers that are used for this purpose now).

The last several paragraphs have described only a few of the possible hardware and software configurations that may be developed in the future. Some of the applications mentioned will be very costly and only groups with large funding will be able to develop them. But that does not mean that all of the future video disc educational applications will be expensive. Some should cost less than existing film or tape programs if enough users are interested. Remember that the cost of an educational innovation must be weighed against the resulting improved instruction. When such improvement occurs, the cost of some innovations likely to develop may seem quite reasonable.

A PROMISING FUTURE

The educational video disc has an established foundation and the potential for a promising future. Although it hasn't been proved successful on any large scale yet, firm commitments by several large corporations and educational organizations demonstrate a faith in the new technology's ability to improve interactive instruction.

Those giants of the home entertainment business that are expanding their operations to include video discs are certainly including educational software in their plans. And RCA, MCA and CBS aren't exactly in business to lose money. They believe the educational video disc has good marketing potential. When IBM, Sony and 3M also decided to enter the video disc business, it became even more obvious that the industrial sector had given the video disc its stamp of approval.

Even though a large number of corporations have entered the business, the market for educational video discs will develop slowly. As previously indicated, much of the hardware isn't even available yet, those players on the market now will probably be improved upon, and development of software has only begun. Writing effective computer programming and evaluating instructional design strategies take time. Even developing the right production techniques to improve a program designed for video disc is a challenge that is not being met overnight. There is still a good deal of research and development that needs to occur before educational video discs will be produced for mass distribution.

"Video disc technology is a results-oriented instruction medium for a results-oriented environment."[5] The video disc for education is now at a point where many educators are becoming aware of its potential. They believe that the video disc "provides a medium that encourages involvement in the learning process—an involvement that brings positive learning results."[6] They now want to get their hands on the players and discs and try it for themselves.

These are exciting times for those working with educational video discs. So many hardware and software applications are being discussed and researched that there's something new almost every day. Trial and error should weed out the unrealistic approaches, and the business of developing programs for distribution will begin. The best applications are still to be discovered, but barring the development of a technology that suddenly renders discs obsolete, the video disc for education, training and interactive instruction is most likely here to stay.

REFERENCES

1. *Videodisc Services of the Nebraska Videodisc Design/Production Group* (Lincoln, NE: University of Nebraska-Lincoln, April 1980), p. 1.

2. One such report available is *Potential Educational Markets for Videodisc Programming: Report on Three Focus Groups,* by Maria Savage and Konrad K. Kalba (Cambridge, MA: Kalba Bowen Associates, Inc., 1978).

3. For more information, contact Dr. George Propp, Media Development Project for the Hearing Impaired, Barkley Memorial Center, University of Nebraska-Lincoln, Lincoln, NE 68501.

4.   Bennion, Junius L., *Authoring Procedures for Interactive Videodisc Procedures—A Manual* (Provo, UT: Division of Instructional Research, Development, and Evaluation, Brigham Young University, March 1976), p. 65.

5.   *Videodisc Services of the Nebraska Videodisc Group,* p. 10.

6.   Ibid.

6

Business Programming
for the Video Disc

by John Rusche

This chapter will offer some insight into the development of optical
video disc software in the business world. It is based on the practical
experience gained by Sandy Corp. in producing more than 30 hours
of video disc programming for 6000 Chevrolet dealerships nationwide.

Sandy Corp. specializes in consultation, instructional design and
implementation of training and communication systems for business
and industry. It uses all communications media, including motion
pictures and video, slides and slidefilm, and audio. For the first time,
the optical video disc allows us to use all our experience in these various media in one coherent program.

Sandy's experience is based on working with the DiscoVision Associates (DVA) model PR-7820 industrial video disc player; however,
most of the remarks below can be applied to future optical video disc
systems as well.

AUDIOVISUAL PROGRAMMING FOR BUSINESS

Audiovisual programming for business and industry is well established, and the video disc is only the latest of a long series of display
devices used for this purpose: slides, filmstrips, film, audio tape,
video tape. There is no doubt that business spends vast sums on A/V
use. *Video in the 80s*, published by Knowledge Industry Publications,
Inc.,* estimates that all organizational users spent $1.1 billion in 1980

---

*Paula Dranov, Louise Moore and Adrienne Hickey, *Video in the 80s,* White
Plains, NY: Knowledge Industry Publications, Inc., 1980.

on video equipment and programming, with business organizations accounting for 60% of the total. Industrial training outlays probably exceed $10 billion annually—no one knows the exact total. Billions more are spent on corporate advertising and public relations. Among the business communications uses to which video discs might be put are:

- employee training;
- management training;
- management communications;
- dealer communications and support;
- marketing and promotion; and
- public relations.

Ford, for example, is replacing its 3/4-inch cassette network with Sony laser optical disc players which will be used at its dealerships for sales and service training, point of purchase programs and management communications.

Sears, Roebuck transferred its summer 1981 catalog onto a laser optical disc in order to test it as an in-store marketing tool at nine outlets in Washington, DC and Cinncinnati, OH. The test, which utilized U.S. Pioneer optical players and DiscoVision discs, was also conducted in 1000 specially selected homes that already possessed video disc players that could play the DiscoVision discs.

For product information and sales or technical training, the optical video disc offers the advantage of flexibility—its ability to provide different types of presentations in a single medium. To the extent that discs are less expensive or easier to use, they will find a place even without the possibility of new types of programming, where they are simply an alternative to slides, film or video tape.

Because early programs produced by the Sandy Corp. tried to take advantage of the features offered by the optical video disc player, we will briefly discuss those that are most applicable to business and industrial programming, particularly in the area of training.

OPERATIONAL FEATURES OF THE OPTICAL DISC PLAYER

As discussed in earlier chapters, the optical video disc system allows the producer to combine moving pictures, silent freeze frame pictures, still pictures with audio over, and two-channel, high quality audio, all under the control of a 1024 (1K)-byte microprocessor, which can be programmed either manually or automatically by pre-programmed video discs. This microprocessor controls such features as freeze frame, frame-by-frame motion, visible picture scan (forward and reverse), slow motion (forward or reverse), individual frame access (five-second maximum search time), frame number display, "selectable" audio, automatic stop on a preselected frame and branching to other microprocessor program steps.

Obviously, the producer must have technical competence in program development for all the above media, plus the system and procedures for bringing them all together on a video disc. Earlier chapters have discussed how material can be created on film or video tape, or taken from existing still pictures and sound tracks, to make a video disc master. In this chapter, further discussion is in order regarding how to use microprocessors in program development.

Microprocessor Control

The combination of a 1024-byte microprocessor and the frame number codes recorded in the vertical interval of the video picture gives the optical video disc its truly unique power.

The video disc functions that are customarily controlled by the microprocessor, either manually or under preprogrammed control, are shown in Table 6.1. Commands that are used to program the microprocessor are shown in Table 6.2.

Table 6.1 Video Disc Functions Controlled by the 1024-byte Microprocessor

1) Audio one and/or two on or off.

2) Slow forward or reverse to a specified frame.

3) Step forward or reverse a specified number of frames.

4) Search to a specified frame.

5) Freeze at a specified frame indefinitely.

6) Play at sound speed and automatically stop on a specified frame.

7) Wait at a specified frame for a specified time.

Table 6.2   **Microprocessor Programming Commands**

Command	Function
1) Program	Places the microprocessor in the "program write" mode.
2) Recall	Calls up a storage register. A register is a temporary memory in which to store frame numbers or other number values.
3) Store	Loads either a frame number or another number into the storage register.
4) Decrement register	Decreases the value stored in the register by one each time it's passed.
5) Input	Prepares the microprocessor to accept a number answer from the user's touch pad.
6) Branch	Instructs the microprocessor "pointer" to go to another program step.
7) Halt	Stops the microprocessor program "pointer."
8) End	Takes the microprocessor out of the "program write" mode.
9) Run	Starts microprocessor program execution at program step zero. If a number is entered before pressing RUN, program execution will start at that point.

To illustrate these functions in practice, we'll develop a very simple program. Remember, after this program is written and checked, it can be recorded on the channel two audio track at the head of every replicated video disc. Then, as the video disc is loaded into the player, the program is automatically "dumped" into the player's microprocessor.

Suppose we had a video disc with descriptive index at frame 151, and a single, five-minute, bilingual motion segment starting at frame 181 and playing to frame 7381. We want the English version on channel one. The last frame of the motion segment contains one review question with one correct and one incorrect answer. If the user answers correctly, we'll bring up a typeset "tell" frame telling him so, and then return to the index. If he answers incorrectly, we'll bring up another "tell" frame informing him and then review a 25-second segment of the film. We'll give him two chances for a correct answer and then return to the index.

The flow diagram in Figure 6.1 illustrates this program. The step-by-step instructions for entering this program into the microprocessor are shown in Table 6.3.

This simple program uses a total of 77 bytes of the available 1024-byte microprocessor memory. It can be entered manually into the video disc player through the use of the touch pad shown in Figure 6.2.

When playing the video disc, the index will automatically be displayed. To play the single motion segment, the user must press 1-7-RUN on the touch pad. The index for the program described

Table 6.3   Microprocessor Programming Instructions

Program Step	Value	Command	Comment
		PROG	Enter the "program write" mode
0	151	SRCH	Search to the index frame and freeze
4	5	RCLL	Call up register 5
6	2	STOR	Load 2 in register 5 for decrement
8	360	STOR	Load reward "tell" frame location
12	370	STOR	Load penalty "tell" frame location
16		HALT	Stop program until next instruction
17	0	AUD2	Turn off channel 2 audio
19	181	SRCH	Locate frame 181
23	7381	ASTP	Play to frame 7381 and freeze
28	7381	SRCH	Search to frame 7381 and freeze
33	3	INPT	Accept user response from touch pad
35	33	BRCH	0 Invalid response
38	47	BRCH	1 Correct response
41	55	BRCH	2 Incorrect response
44	33	BRCH	3+ Invalid response
47	6	RCLL	Recall register 6
49		SRCH	Search to reward "tell" frame
50	30	WAIT	Freeze reward frame 3 seconds
53	0	BRCH	Return to index frame
55	7	RCLL	Recall register 7
57		SRCH	Search to penalty "tell" frame
58	30	WAIT	Freeze penalty frame 3 seconds
61	1520	SRCH	Locate review start frame 1520
66	2120	ASTP	Play review section to frame 2120
71	5	DECR	Subtract 1 from register 5
73	28	BRCH	Go to quiz if register 5 > 0
76	0	BRCH	Go to index if register 5 = 0
		END	Exit the "program write" mode

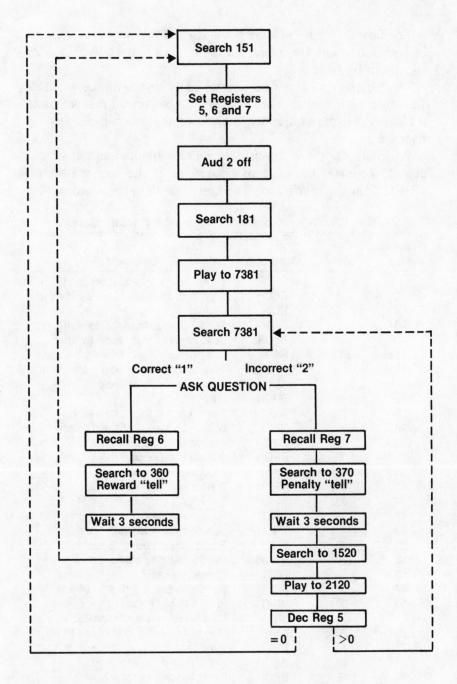

Figure 6.1 Flow diagram of simple video disc program.

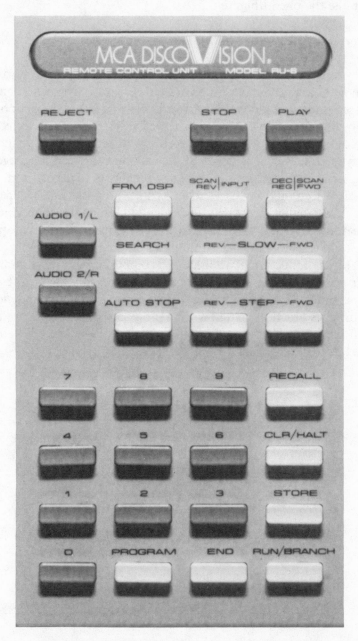

Figure 6.2   Video disc player remote touch pad.

above is extremely simple. More typically an index would look something like the one in Figure 6.3.

Microprocessor Program Variations

Obviously, there can be a far greater variety of program-controlled movements than those illustrated in Table 6.3. More segments can be accessed, more questions can be asked, and there can be more possible answers or a greater variety of remedial instruction connected to these questions. For bilingual video discs, different sets of graphics and type frames can be accessed, depending on which language is being used. Finally, the microprocessor permits us to depart from the necessity of cutting sequences together in continuity. The sequences can be cut together in virtually any order and, as long as about one or two seconds of black is permissible between sequences, the microprocessor can assemble various continuities on command.

To assist production of accurate and complex programs, the Sandy Corp. has developed, and is continuing to experiment with, off-line microcomputer systems. The link between microcomputer and video disc presents practical help in writing programs such as the one illustrated here, as well as some exciting possibilities which will be discussed later in this chapter.

Detail Index
New Product Presentation

10	• Table des Matières en Français
100	• Full Product presentation
125	• Models
150	• Interior Comfort
175	• Ride and Handling
200	• Powertrains
225	• Built-in Values
250	• Options
RUN/INDEX	Return To Main Index

To view these programs, press index number followed by the "RUN/INDEX" button on the touch pad.

Figure 6.3 Sample video disc index frame.

Figure 6.4 shows the microcomputer-to-video disc link developed by Sandy Corp. At left is the video disc player connected to the microcomputer by way of a proprietary "black box." A check video disc is seen on the TV screen. At right is a 3/4-inch video cassette player connected to a frame-accurate shuttle device. A check video cassette with visible frame number is seen on the TV screen. A close-up of the microcomputer-to-video disc link is shown in Figure 6.5. The GM Video Center is shown in Figure 6.6

PRODUCING VIDEO DISCS FOR CHEVROLET

When we first learned we would be releasing our Chevrolet in-dealership communications and training programs on video disc, there were no production disc players in existence. There were no technical manuals. There were no instruction manuals. There were no defined systems or procedures. And the assignment was to deliver original video disc programming to 6000 Chevrolet dealerships within nine months.

In the beginning, Sandy Corp. was faced with an enormous job of planning and program design. For example, while instructional designers, writers, and production people were busy in the actual development of the video disc software, others at Sandy had to concern themselves with the fact that some 30,000 Chevrolet salesmen and 60,000 technicians in dealerships would be using these discs on hardware they had never before seen or operated. So, among other training activities, a simple "follow-the-numbers" instructional diagram was developed to coach users of the video disc hardware and ensure maximum use of the all-important software. (See Figure 6.7.)

Note that in Figure 6.7 there are fewer buttons on the touch pad control than in Figure 6.2. This is accomplished by fitting a plastic template over the touch pad buttons to mask all but those buttons needed for actual operation of the video disc player in a dealership environment. This makes operation easier when using the equipment with customers.

Our primary focus for original video disc production was about two and one-half hours of programming designed for point-of-sale advertising and sales training for the 1980 Chevrolet passenger car and truck lines.

The development cycle for video discs of this type is lengthy and

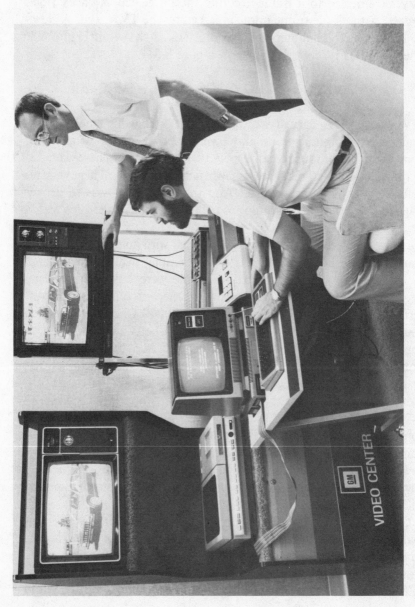

Figure 6.4 Sandy Corp. microcomputer-to-video disc link being used to finalize a video disc program.

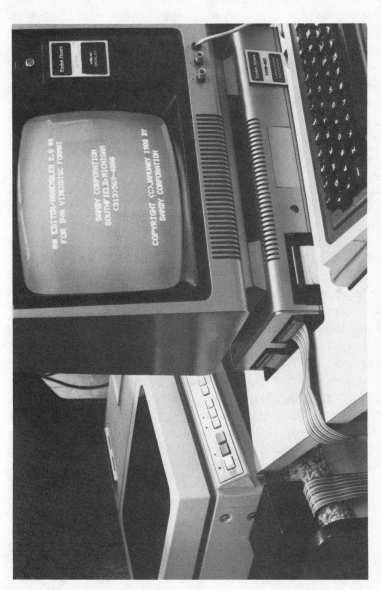

Figure 6.5 Close-up of the Sandy Corp. microcomputer-to-video disc link used to develop microprocessor programs and provide off-line control of the video disc player.

must be flexible until virtually the last minute. Several months were required to complete the process of research, script development, photography, finishing and video disc mastering/replication. The video discs had to depict the products accurately from an engineering, legal and merchandising point of view. Because they were very complex, we had to start early but be ready for many last-minute changes in products and marketing strategy that would affect the finished discs.

Each of the five video disc sides contains about 25 minutes of motion picture shot and finished on 16mm film. The sequence for each product line can be accessed and shown in its entirety or, by referring to the index, shorter segments on such topics as interior comfort features, powertrains and cargo capacity can be shown. These motion sequences have an accompanying sound track in both English and Spanish.

In addition, each side contains about 60 silent still pictures consisting of art, photos and type. These still pictures show certain features and benefits and serve as the centerpiece for a sales meeting or as a discussion aid with customers. The stills can be accessed and controlled in much the same manner as a slide projector.

Finally, each video disc side contains about 10 multiple-choice review questions. These questions are programmed to provide a positive or negative reinforcement frame, followed by remedial instruction for an incorrect answer.

As the design of these video discs began taking shape, it became apparent that the 1024-byte microprocessor would not hold enough programming to execute the required instructions. Thus, the "double-dump" was created. As noted earlier, the microprocessor program for preprogrammed video discs is carried on audio channel two at the very beginning of the disc and dumped into the microprocessor when loading. We discovered that additional programming dumps of 1K bytes each could be tucked away in other parts of a video disc as long as there is no content audio on channel two, and preferably no picture. These "double dumps" permit additional programming capacity, but they cannot interact with each other.

Figure 6.6   The GM Video Center being used by John Rusche. The system includes a Zenith System 3, 19″ color TV receiver, a DVA industrial video disc player model PR-7820 and a remote touch pad control model RU-8, all enclosed in a custom-built moveable cabinet.

THE VIDEO DISC DEVELOPMENT PROCESS

The complex process required to create a video disc with microprocessor preprogramming, from research and script through replication and distribution, includes the following main activities:

1) Client, sponsor or expert input and approvals.

2) Research, script development, storyboards.

3) Production and finishing in film or video tape, motion and/or stills.

4) Foreign-language translation, production and finishing of graphics and audio.

5) Microprocessor program development.

6) Video disc premastering, mastering and replication.

7) Development of related graphic materials such as video disc labels, video disc jackets and supporting reference booklets or training guides.

Let's look at each of the activities in a little more detail.

1. *Client, sponsor or expert input and approvals.* This is very important at the beginning of a project, assuming that the producer understands media production and the client understands his product. They must get together to exchange views and define objectives. Depending on the content of the video disc, this input can be sought from many different perspectives: design, engineering, legal, marketing, merchandising, etc. It's required at various scripting stages, as well as at the workprint, answerprint and check video disc stages.

2. *Research, script development, storyboards.* These phases are basically the same as in writing for more conventional media, with one very important addition—video disc program logic. The writer must envision how the power of the video disc will be used and com-

municate that to the production and programming people. A good place to start is by writing a "working index" early in the script stage: the writer should consider what the index will look like when it comes up on the TV screen at the head of the video disc. As the scripting develops, both the index and the program logic must be constantly kept in mind so that the microprocessor programming described in step 5, below, will be much easier. (See Chapter 5 for a more detailed discussion of scripting.)

3. *Production and finishing in film or video tape, motion and/or stills.* In most respects, this simply requires production with good television practices in mind. Toward the end of the production cycle, after the workprint approval, the video disc indices and tell frames are typeset and photographed. All film and sound tracks are assembled according to technical procedures established by Disco-Vision Associates for transfer to video disc.

4. *Foreign-language translation, production and finishing of graphics and audio.* After workprint approval, a U-matic video cassette, plus various dubs of the English soundtrack, are made. Script translation is completed, and all recording and graphics are produced. The music and effects soundtrack from the English version is mixed with the foreign-language voicetrack. The final mix is then transferred to the second channel of a 1-inch helical, Type C video tape or channel three on 35mm sprocketed, magnetic stock.

5. *Microprocessor program development.* Based on the logic provided in the script, a variation of the flow diagram in Figure 6.1 is developed and refined. At the workprint approval stage, precise frame counts of various sequences are taken. These are then used to write a program similar to the one in Table 6.3. Depending on the complexity of this program, it is then verified against a U-matic or a check video disc (see Figure 6.4) at a later stage in the process, and any necessary revisions are made.

6. *Video disc premastering, mastering and replication.* This process was described in earlier chapters. It should be noted that as time passes, many of the processes in this phase could change with increased automation, new inventions and competitive pressures.

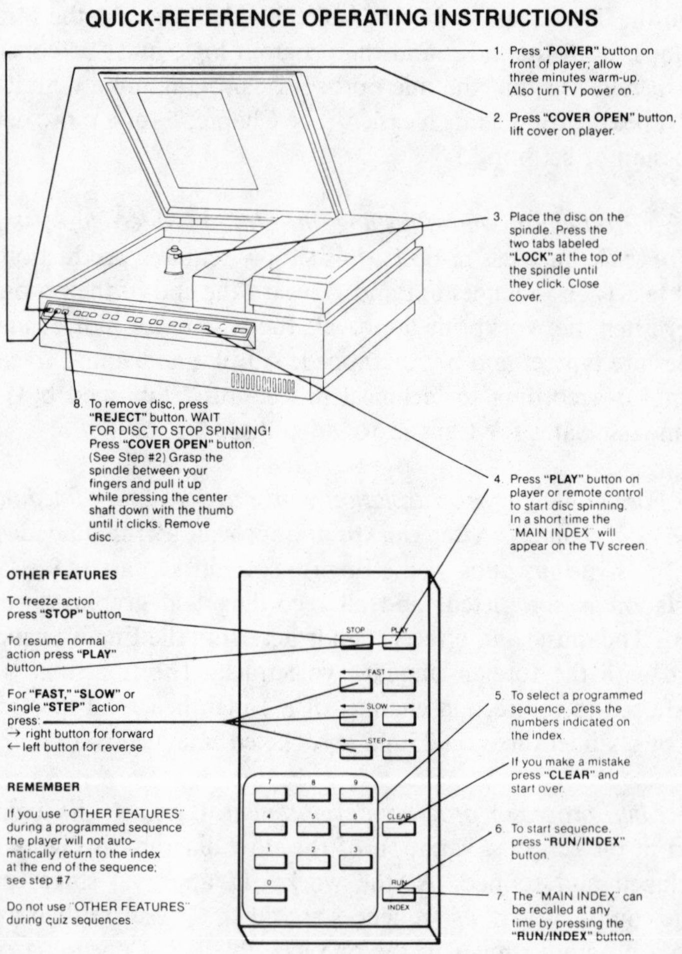
QUICK-REFERENCE OPERATING INSTRUCTIONS
1. Press "POWER" button on front of player; allow three minutes warm-up. Also turn TV power on.
2. Press "COVER OPEN" button, lift cover on player.
3. Place the disc on the spindle. Press the two tabs labeled "LOCK" at the top of the spindle until they click. Close cover.
8. To remove disc, press "REJECT" button. WAIT FOR DISC TO STOP SPINNING! Press "COVER OPEN" button. (See Step #2) Grasp the spindle between your fingers and pull it up while pressing the center shaft down with the thumb until it clicks. Remove disc.
4. Press "PLAY" button on player or remote control to start disc spinning. In a short time the "MAIN INDEX" will appear on the TV screen.
OTHER FEATURES
To freeze action press "STOP" button
To resume normal action press "PLAY" button
For "FAST," "SLOW" or single "STEP" action press:
→ right button for forward
← left button for reverse
REMEMBER
If you use "OTHER FEATURES" during a programmed sequence the player will not automatically return to the index at the end of the sequence; see step #7.
Do not use "OTHER FEATURES" during quiz sequences.
STOP
PLAY
FAST
SLOW
STEP
7
8
9
4
5
6
1
2
3
0
CLEAR
RUN
INDEX
5. To select a programmed sequence, press the numbers indicated on the index.
If you make a mistake press "CLEAR" and start over.
6. To start sequence, press "RUN/INDEX" button.
7. The "MAIN INDEX" can be recalled at any time by pressing the "RUN/INDEX" button.

Figure 6.7 Quick-reference video disc player operating instructions.

7. *Development of related graphic materials.* As with any audio-visual training program, the appeal and instructional value of the materials are heavily influenced by the supporting graphic materials. Customarily, Sandy Corp. produces video disc jackets containing a summary of the index, a content overview and a brief review of how to access the programs contained on the disc. In addition, supporting training guides, study guides, meeting leader's notes, reference booklets, job aids and/or instructional posters are usually developed.

APPLICATIONS

Altogether, in the first year of production for General Motors, Sandy Corp. developed more than 30 hours of programming that included video discs designed for:

- Salesperson new product training and point-of-sale customer viewing.

- Salesperson motivation and selling skills training.

- Service technician product training.

- Service advisor skills and technical training.

Within each of these categories, there is broad variation in terms of production technique employed and complexity of microprocessor programming. Some are composed primarily of still pictures with audio-over, others are primarily motion pictures.

The dual-language video discs also make very effective learning tools for foreign-language students, because they can check their understanding of the language used in the program by simply pressing a button to bring up the English version of the scene.

Product or service training video discs can cover whole lines or models of products in a very cost-efficient manner. The features that are generic can be placed on one portion of the disc and the features that are unique to the different lines on another portion. The microprocessor program can then pull these various elements together in a variety of presentations.

VIDEO DISC APPLICATIONS OF THE FUTURE

As video discs become more widely available, there is practically no limit to the business applications.

A single video disc can combine motion, sound slide stills, silent slide stills, text stills (like a book), two-channel audio and microprocessor control. As we all know, the potential for creative use of any one of these media by itself is significant, but when they can all be combined in one densely packaged form, the potential is limited only by experience and imagination.

At this point, the single most limiting factor for truly sophisticated video disc programs is the capacity of the 1K-byte microprocessor. This can be overcome in the future by either upgrading the built-in microprocessor, or by connecting the video disc player to an external micro- or minicomputer.

In this way, it's possible to use the external computer's floppy disc to store and generate different versions of the video disc control program. In addition, this floppy disc can contain different types of reader frames, questions and solutions which are generated on command by the computer, not the video disc, and appear on the TV screen.

Generic and/or unchanging visual information, such as military or business operations, symptoms of disease, automotive mechanical problems, etc., can be recorded on the video disc. These problems or methods may change over time, e.g., with the availability of new military or business hardware or systems, new drugs or treatments, new mechanical test equipment or replacement components. As these changes occur, it would be much less costly to modify material stored on the computer floppy disc than it would be to produce new video discs. Clearly though, to make development of highly sophisticated, interactive video disc programs cost effective, the subject should be one with a relatively long "shelf life."

Linking a computer to a video disc player also opens up other possibilities. The use of videotext is spreading around the world. Videotext transmissions combine textual information, both words and numbers, on a TV screen. Their main drawback is lack of ability to integrate photographs or motion pictures. On a private network basis, it should be feasible to combine the features of videotext with video disc visuals.

Still another point-of-sale possibility is to use the computer and the TV screen as a prompting device for customers. The screen could ask a lead-in question and request the customer to respond by touching the screen with his hand or a light pen. In this way the computer could first lead the customer through a series of qualifying questions; then demonstrate the product on video disc in full color and sound; and finally, perhaps, even close the sale.

Many links between computer systems and visual programming on the video disc are likely to emerge as users gain experience with the new technology. Consider, for example, the travel information and reservation system that American Express is designing for its travel agents. Some 150 disc players were scheduled to be placed in agents' offices in late 1980 as part of an ambitious system known as TRIPS (travel information processing system). By using terminals linked to the system, agents will be able to check on available reservations from airlines, hotels and car rental companies. They can also make reservations and automatically print out tickets using the system. What the disc players will add is a unique enhancement: discs will display color photos of hotel rooms or resort locations. After viewing, the traveler can make his choice, and the agent can place and verify the reservation on the spot.

Ultimately, the TRIPS system could be in 1000 travel offices affiliated with American Express around the world. According to American Express officials, the random access capabilities of the optical disc are an essential reason for its selection.

Yet another example of how a computer-controlled disc player might be used in marketing comes from On-Line Media, Inc., which began, in May 1980, to put TV sets and video cassette recorders in supermarkets to show commercials to shoppers. While the initial system uses VCRs, a later version, being developed with DiscoVision Associates, will use computer-controlled disc players that can be activated over telephone lines.

In any event, it takes significant time and money to create programs for any of these possible uses. This leads us quite naturally to a very basic question.

ECONOMICS OF BUSINESS PROGRAMMING
ON VIDEO DISC

As usual, while this technology is in its youth, the most interested users are business/industrial, military and educational organizations —those with the most to gain if they can apply the technology to advantage. The question is how to evaluate this new technology in terms of its practical application to learning and communications, and its cost effectiveness for these applications.

Creating straightforward visual programs on video disc should be comparable to the cost on film or tape, perhaps even less if still frames can be used in place of certain motion sequences. But there is no getting around the fact that creating original programming that takes advantage of the optical disc's advanced features is expensive. Using microprocessor programming to do branching or interactive teaching calls for a high level of skill in program creation, as well as added time in scripting and shooting. If standard costs for creating business films or tapes can range from a few hundred to a few thousand dollars per minute, the cost for a minute of video disc programming could be three or four times that amount. And these costs are before the expense of mastering and replication. Mastering is more expensive for discs than for tape, while replication costs are higher for discs until the quantities duplicated become substantial. Thus, the decision to produce video discs must be made on the basis of a careful analysis of the needs of the viewers, and the nature of the material to be created.

As an example, is it important that users be able to record? There is no practical capability of this with the video disc. In some cases, lack of recording ability may be an advantage—particularly in applications where uniformity of message is desired and careless use by nonprofessionals may cause unwanted erasures of valuable material. At present, the video disc is not a do-it-yourself medium.

How do you plan to use still frames? Their availability is a major potential advantage over video tape, since any frame can be found and held on the optical disc without distortion. But you cannot use still frames on the disc with narration, unless the frame is animated to cover the narration.

How many duplicates will you usually need? As noted above, video discs are generally not cost effective for very small numbers of dupes. However, market pressures have recently driven video disc costs down, and may continue to do so.

In the first year, DiscoVision Associates' published industrial video disc prices for a one-hour show, without microprocessor programming, required you to order 1000 video discs before the cost per disc equaled the cost of video cassettes in the 1/2-inch format. For cassettes in the 3/4-inch format the cost crossover point was about 200 video discs. In the summer of 1980 DiscoVision Associates cut these published industrial video disc prices by 40% to 60%. This brought the cost crossover point for a one-hour show down to 200 and 100 copies when we compare discs to 1/2-inch and 3/4-inch video cassettes, respectively.

Of course, when microprocessor programming is added to video discs they are relatively more expensive than video cassettes, but then you also gain capabilities unavailable with video cassettes.

It is entirely possible, as others enter the video disc replication field, and as technology and manufacturing processes improve, that these prices could be driven down even further.

There is no question that the video disc is an intriguing and versatile training and communications tool. For the first time ever, it offers the opportunity to select the proper media mix to satisfy specific business training and communications objectives, and to combine them on one medium.

The power of the microprocessor adds a new dimension to visual programming, and it will take time and experimentation before we can begin to realize its potential. But if there is an institutional market for video discs in general and the optical disc in particular, business users should lead the way. Already they have a large and growing investment in video as a communications tool. They also have the financial incentive to spend money on technologies that will improve the performance of their employees, since improved performance will mean improved profits. More and more companies are finding that communications—in the form of advertising, management and employee motivation, training, customer service, public relations—is central to their success. There is every hope that the video disc will be a part of the evolving world of business communications.

7

Video Discs for Information Storage and Retrieval

by Alan Horder

Whereas in its application in the consumer market the prime purpose of the video disc is to provide *moving* pictures, in information storage applications the disc is required primarily to provide *still* pictures. The future of video disc systems making extensive use of the still-picture facility is likely to lie with *optical* video disc systems. The present chapter of this book is therefore largely concerned with optical video discs.

The main features of the optical video disc that make it a potentially attractive medium for information storage and retrieval are:

- high information storage capacity

- low information storage cost

- rapid random access

INFORMATION STORAGE CAPACITY

Information can be encoded on video discs in a number of ways. In considering the information storage capacity of such discs we need to distinguish in particular between discs designed to be television-

This chapter originally appeared in slightly different form in Horder, Alan, *Video Discs—Their Application to Information Storage and Retrieval*, Bayfordbury, Hertford, Hertfordshire, England: National Reprographic Centre for documentation, 1979.

compatible and discs designed to be computer-compatible. The advantage of using a television-compatible encoding system is that discs can be produced using mastering techniques developed for the production of consumer video discs. This has the advantage of minimizing production costs and also means that the discs can be played on consumer as well as industrial players. If, however, video discs are to be used as computer stores, a computer-compatible coding system must of necessity be employed. Although this means that such discs can no longer be played on consumer players, use of a binary information coding system—used universally in computer storage and transmission—leads to a greatly increased storage capacity for the disc.

We can gain a general picture of the information storage capacity of a television-compatible optical video disc by taking the Disco-Vision disc as an example. Each side of this disc has a maximum of 54,000 tracks which, in the still-frame mode, yield up to 54,000 frames. A rough calculation shows that the area occupied by a single frame on this disc, which has a diameter of 305mm, is approximately $1mm^2$. (If the disc is double-sided we can halve this to $0.5mm^2$.) This compares with the $160mm^2$ occupied by a single frame on a 98-frame microfiche or the $5mm^2$ on a 3000-frame ultramicrofiche.

Even when encoded with a television-compatible signal, therefore, the video disc would appear to have a much higher information storage capacity than microfiche or even ultramicrofiche. Before pressing this comparison too far, however, we need to recognize that the amount of *text* that can be contained in a single frame in the different media and remain legible is not necessarily the same. The amount of text that can be legibly displayed on an ordinary television screen is, in fact, quite limited—in particular by the line structure of the raster scan that forms the image. It has been stated that when displaying textual material on a 625-line video system a maximum of 1500 characters can be legibly displayed on the screen, and it is of interest to note that the standard format adopted for teletext (viewdata) systems provides for a maximum of only 960 characters in the display. This compares with a maximum of about 4000 characters on a typed page (8¼ inches x 11¾ inches) or 10,000 characters on the printed page of the NRCd journal "Reprographics Quarterly." An estimate relating to a 525-line television system gives a typewritten U.S. page (8½ inches x 11 inches) as requiring *eight* normal television frames.

Whereas materials specially produced for display from a video disc can be formatted with this limitation of the television display system in mind, it will be apparent that a number of problems must be solved before many types of *existing* materials can be stored on a video disc. It may be possible to record some materials in sections and to view these sequentially—a practice sometimes adopted in the microfilming of very large drawings. With many types of material, however, adoption of such a "matrix technique" would pose unacceptable problems on retrieval. In the latter case the solution may be to use a special high-definition television monitor instead of the usual domestic receiver. Whereas the screen of a domestic television has a 625 or 525 scan line screen, the screen on a high-definition monitor may have up to 2000 lines. Such a screen will be capable of giving a legible display of most types of printed materials. Consideration has already been given to the development of "scan conversion" techniques that will enable video discs produced to normal television standards to feed such high-definition monitors.

As noted above, information on video discs intended for use as computer stores must be recorded not in a video (analog) format but in the way that is universally used for computer storage and transmission, i.e., in a binary coded digital form. In considering the information storage capacity of a video disc used in this mode it may help if we again take the DiscoVision disc as an example. The storage capacity of a single side of this is quoted as 10^{10} bits. This equates to 1250 million 8-bit characters or 1 million 1250-character pages. This is obviously very much greater than the capacity of the same disc encoded to produce a television-compatible raster, which we saw was some 54,000 pages, each capable of displaying a very similar number of characters.

As another example of the storage capacity of a video disc used as a computer store we may take the Digital Recording system, the application of which in the area of mass data storage is now under study. The storage capacity claimed for this is 3×10^8 bits per square inch, giving an information capacity on a 4 inch x 5 inch rectangular "disc" (videofiche?) of 6×10^9 bits—very similar to that claimed for a single side of a DiscoVision disc. Using suitable coding techniques an information density "some 7 to 8 thousand times more dense than microfilm" is claimed for the Digital system. Developments in technology may be expected to increase the storage capacity of video discs still further in time. The chief scientist of the Xerox Corp. predicted in 1978 that "by the mid-1980s . . . the entire contents of the 18 million volumes in the Library of Congress will be able to be stored on 100 optical video discs." The basis for this prediction is not stated but it would appear to imply a packing density several orders greater than has been achieved on production video discs so far. Time will tell whether this can be achieved but it needs to be recognized that the storage capacity of optical video discs is limited ultimately by the wavelength of light.

The future of video discs as mass memories obviously lies with systems providing very high information packing densities. One way of increasing this is to encode the information even more efficiently. The development of optimum coding schemes is therefore currently being studied in order to exploit the bit-packing density of video discs to the full.

INFORMATION STORAGE COSTS

Any figures for information storage costs with respect to video discs quoted at the present time can only be tentative and will obviously depend on the number of copies made of the master disc. A cost per bit of 10^{-5} (U.S.) cents was quoted for the DiscoVision disc in 1976. A more recent estimate is that a cost as low as 10^{-8} cents per bit stored on video disc should be achievable.

RAPID RANDOM ACCESS

Just as access to an individual track on an ordinary phonograph record can be faster than to a chosen part of an audio cassette recording, and access to an individual frame of a microfiche faster than to a single frame on a roll of microfilm, so access to an individual track on a video disc can be faster than to a selected part of a video tape. Rapid random access is, therefore, an important advantage of the video disc as an information storage medium.

In the DiscoVision Industrial Player the average time for retrieval of any one of the 54,000 frames of information on the disc is 2.5 seconds; the longest time is five seconds. A random access time not exceeding two seconds is claimed for the Thomson-CSF player. While these times match those achieved with information storage systems employing microforms, they are long compared with the access times (measured in microseconds) commonly encountered in computer storage. There is, however, evidence that the presently achieved access times can be reduced.

DIRECT-READ-AFTER-WRITE SYSTEMS

The video disc systems now being readied for the consumer market are all read-only systems, employing prerecorded video discs produced in quantity using similar pressing methods to those employed in the manufacture of phonograph records. The use of such production methods is central to the economics of consumer video discs. The consequent absence of a recording facility on consumer video disc players is not seen as a major disadvantage. Among software providers it is even welcomed.

The situation is not necessarily the same when we come to consider information storage and retrieval applications of video discs, although here we must distinguish between what we may call "library applications" on the one hand and "business applications" on the other. The greater part of the collection in the average library consists of printed publications, produced in multiple runs using conventional printing processes. The read-only type of discs would appear quite suitable for the storage of many types of materials in this category.

With business records, however, the situation is different. The greater portion of these are produced either locally within the company holding the records or in companies it deals with. Furthermore, records of this type are produced either as single copies or in sets of not more than a handful of copies—certainly not usually in the multiple runs typical of book or periodical publishing. Many of the records require updating from time to time. In this situation a video disc system with which the user can "write" as well as "read" is essential. Further, the system must permit the economic production of *single* copies of discs.

A number of companies have recognized this need and are working to produce equipment and materials to meet it. Among the companies working in this area is Philips, which has developed a DRAW (*d*irect-*r*ead-*a*fter-*w*rite) system based on the VLP consumer disc. Work on this development has been partly supported by a U.S. Defense Advanced Research Projects Agency (DARPA) contract. A tellurium-based coating has been selected as the recording material. Other companies reported to be working in this field include Robert Bosch (West Germany), IBM (U.S.), RCA (U.S.), Hitachi (Japan) and Matsushita (Japan). The Drexler Technology Corp. has developed a special recording material for use in such systems. Named DREXON™, this reflective plastic/metallic composite records instantly without processing of any kind.

PRINTOUT FROM VIDEO DISCS

Hard copy of an image appearing on a television screen or video display unit can be produced in a number of ways. The screen can be photographed using, for example, an ordinary camera and Polaroid film; copies produced in this way will, however, be comparatively expensive. Some video display units (e.g., Tektronix) have a built-in printer unit and are thus analogous to microfilm reader-printers. More interesting possibilities for printout from computer-compatible video discs are opened up by laser printers developed for high-speed production of hard copy from computers. Printers such as the IBM 3800, Xerox 9700, Canon LBP 3500 and Siemens ND-2 can print at page rates up to two pages per second—the speed of the fastest machines equaling that of computer output microfilmers.

THE FUTURE OF VIDEO DISCS IN INFORMATION STORAGE AND RETRIEVAL

If we can learn anything at all from the history of their development to date it is that the future of video discs is unpredictable. This applies to the technology, the timetable and the interaction between the two. Any observations here about likely developments in the application of video discs in information storage and retrieval can therefore only be general.

In the early development of video discs the future of industrial systems seemed to be inextricably linked with the future of consumer systems. This is probably still the case with television-compatible models. Computer-compatible video disc systems would, however, now appear to have a future of their own, irrespective of the success or failure of video discs in the consumer market. This would seem to apply particularly to systems with direct-read-after-write capability.

One area of video disc technology peculiar to the use of video discs in information storage that would appear to require development is that of the inputting of material. *New* material can be formatted by key strokes on a variety of types of equipment—in particular, on word processors. The inputting of *existing* printed material, however, presents quite different problems. Documents for inputting can, of course, be simply placed before a television camera, but the use of scanning techniques or even OCR techniques may be more appropriate in many situations. Aspects of the problem that will require special attention will be image quality and economics.

There remain, of course, many unknowns. It is one thing to speak of putting the whole of the 18 million volumes in the Library of Congress on 100 video discs and quite another to be sure that users will be content to consult information stored in this form! We hear enough about the problem of "user resistance to microforms." Are we merely going to exchange this for "user resistance to video discs"? Only time will tell. It could, however, be that the greatly increased ease of access to information provided by computer-assisted retrieval which the video disc facilitates, coupled with the control of image quality possible on an electronic display device but lacking on a microform reader, will provide the user with sufficient "value added" to overcome his or her resistance. We should be well on the way to knowing if this is the case by the mid-1980s.

This problem will not, of course, arise in applications of the video disc in computer storage, where the user has no means of knowing—and is not concerned to know—whether the information viewed on a video display unit emanates from a solid-state memory, a floppy disc or a video disc.

In conclusion, then, it would appear that in the video disc we have a medium that within the next decade will come to pose a real challenge to microforms. This challenge will probably be felt most in applications where remote access to information is required. A number of systems have been developed and others are currently in the course of development which will give remote access to microforms by means of some form of television link. Some of these systems propose to use coaxial cables to transmit information; others public telephone lines. But whatever the form of transmission used, all involve the conversion of a human-readable image on a microform into an electrical signal. A video disc system, in which the initial signal from the information store is an electrical one, would appear to have advantages in this situation, especially if it eliminates the need for the mechanical transport of the storage medium that characterizes large-scale microform stores.

REFERENCES

Kenney, G.C., "Special-Purpose Applications of the Optical Videodisc System," *IEEE Transactions on Consumer Electronics,* 1976, CE-22(4): 327-338.

Turner, I., "The Use of Video Communication Systems in Information Retrieval," *Reprographics Quarterly,* 1974, 7(2): 48-54.

Dyall, W.T., "Televiewers: What Do You Mean by High Resolution?" *Electro-Optical Systems Design*, 1978, 10(3): 26-29.

Digital Recording: An Optical System for High Density Information Storage and Retrieval, Digital Recording Corporation, Wilton, CT (undated): 26pp.

Ammon, G.J. et al., "Recording by Optical Video Disc," *Systems International*, June 1978: 32-33.

Bartolini, R.A. et al., "Optical Disc Systems Emerge," *IEEE Spectrum*, 1978,15(8): 20-28.

Kenney, G.C. et al., "An Optical Disc Replaces 25 Mag Tapes," *IEEE Spectrum*, 1979, 16(2): 33-38.

Drexler, J., "Drexon (TM) Disc Storage for Laser Printers," paper presented at SPIE Laser Printing Conference, Los Angeles, CA, January 1979.

8

The Video Disc and Competing Technologies

by Mark Schubin

Virtually all the video disc systems described in this book had their origins in research begun in the 1960s, when video tape recording was a process that required expensive machinery and consumed a great deal of tape.

Consumer television, in that period, consisted exclusively of television sets on which viewers could watch three national networks and, perhaps, a few local independent or educational stations in some markets. There were few cable television systems, serving less than 1 million homes, and even these were simply providing improved reception of the few broadcast signals. Almost no educational or business organizations had their own networks or video tape players.

There was no such thing as a video game and satellites were used only to transmit an occasional network news story from overseas.

It was into this virtually empty market that the video disc system developers hoped to introduce a new source of television programming for homes and institutions. Technology continued to advance, however. Today, in addition to home video tape recorders and video games, there are even home computers. Fifteen million homes have access to multiple channels of cable television programming, including satellite feeds of sports and recent movies. Informational services offering access to data banks are beginning to be available with information transmitted via television signals or by telephone; in some parts of the world, satellites are beginning to transmit directly to home television sets. Many thousands of business, medical and educational institutions create their own video programming and distribute it on VTRs, or even via satellite.

TECHNOLOGIES FAVORABLE TO THE VIDEO DISC

Despite all of the new sources of television programming for homes and institutions, some of the recent technological advances have actually opened new markets for the video disc. Primary among these is the information revolution.

The video disc is an extremely high density information storage medium. A disc that can hold an hour of television programming is actually storing 108,000 successive television frames. Moreover, each frame consists of two successive fields, each containing all of the odd or all of the even scanning lines. Since individual fields can present acceptable pictures, the capacity of an hour-long disc is 216,000 color pictures. However, until a breakthrough in the price of field storage occurs—as will shortly be discussed—there is no way of separating those fields from the frames or, for that matter, of separating any pictorial information of shorter duration than a single rotation of the disc. Therefore, for each hour of recorded information, an 1800 rpm disc system can offer 108,000 pictures, a 900 rpm disc system can offer 54,000 pictures, and a 450 rpm disc system can offer 27,000 pictures. (It is actually better to refer to these as pictures than as frames, since the 27,000 pictures of a 450 rpm system each consist of four inseparable frames.)

The 54,000 pictures available on a 900 rpm disc or on a half-hour side of an 1800 rpm disc are comparable to the number of words defined in an average paperbound dictionary. Thus, a dictionary disc might be produced in which each word could have a defining or descriptive picture (though just how to illustrate such words as "gentility" or "stench" pictorially may prove a problem). Of course, putting this information on the disc is no easy matter, since each picture must be shot and stored (probably on video tape) and then edited into the succession of pictures required. If each shot is recorded for just five seconds (which might prove an editorial nightmare, utilizing today's video equipment), a 108,000 picture disc would require 150 hours of video tape. Then, too, there is the question of the price to place upon such a disc.

More realistically, the pictorial capability of the disc could be used for training programs, where motion sequences might be followed by one or more still pictures to be studied at the trainee's leisure, and then be followed by more video information. In this fashion, many hours of instructional material might be placed on a single disc. Alternatively, the disc might contain programmed instruction: motion sequences followed by alternative still frames and/or motion segments, depending on the trainee's selection. Thus, an incorrect answer to a question might bring remedial information, or, as a preventive measure against ennui, a series of tangents might be provided, all eventually covering the appropriate material. Various ways of creating such instructional video disc programs are discussed by Paul Merrill in Chapter 4.

Microprocessors and Picture Coding

The keys to the utilization of video disc systems in this fashion are microprocessors and picture coding, both fairly recent technologies. The first microprocessor was only available as recently as mid-1971, and was developed as a tiny, general purpose computer on a single semiconductor chip less than 1/60 of a square inch in diameter. Originally costing hundreds of dollars, these tiny computers eventually fell so low in price that, to illustrate this price drop at a 1978 technical meeting, a representative of Fairchild Camera and Instrument flung a pocketful of microprocessors into the audience.

Controlling a sophisticated video disc player without the benefit of a microprocessor can be a horror. The usual functions, such as "play" and "stop," are easily implemented with ordinary switches. A variable speed slow motion is somewhat more difficult, but might also be achieved utilizing conventional control technology. Frame-by-frame picture advance is still harder, since a tiny advance (on the order of 60 billionths of an inch) is required; and the fact that the scanner, or head, has left one track and is on the next must be recognized, amidst confusing error signals. Random access could be even harder, if the circuitry had to keep track of how many of the 54,000 frames had been passed.

Yet it is fairly easy to add coding to a particular frame or track, and to allow a microprocessor to read this coding to determine location. Random access is simply a matter of locating the appropriate code; frame-by-frame advance is simply a matter of finding the next coded number. In fact, only 16 digital bits, or pulses, are required to identify more than 65,000 frames or pictures.

However, once a microprocessor is used to handle these relatively simple functions, it can perform more sophisticated tasks as well. A microprocessor controller might easily select and repeat various sections of a disc to effectively loop a section of action or music. For example, to show a lengthy segment on the repetitive operation of a machine tool, one might record one complete cycle of the operation and instruct the microprocessor to repeat that cycle for the desired duration. Various views of an animated character might be selected by the microprocessor to perform its own animation sequence. For such animation sequences, the ability to change frames without marked flickering would be important—a capability the video disc system does not presently have in the still frame mode. A slight flicker at the frame shifts might be acceptable, and future storage devices could alleviate the problem entirely.

A built-in microprocessor of a video disc system can also communicate to external microprocessors, such as home computers, to make programming and interaction easier. Thus, a home computer that uses a television set as its display device might instruct the video disc to feed certain images or action sequences to which the computer could then add notes, arrows and/or instructions. Each response by the viewer could instantly change the action and notes.

These are only a few of the possibilities that microprocessor-controlled video disc players could offer. Some disc players are being designed with built-in computer connections for easier interaction with home and industrial computers. Connected to a computer, however, a video disc system is not limited to pictorial and aural information storage.

Dealing with High Output Rates

The coding scheme used by most video disc systems could allow the storage of approximately 36 billion bits of raw data on each one-hour disc, which could be recorded and recovered at the astounding rate of roughly 10 million bits per second. No home computer and few large systems can deal with such a rate. However, small digital memories, unlike pictorial memories, are fortunately very inexpensive and readily available, and can buffer this high output to a rate that might be accepted by a home computer. Still, the problems of assembling 36 billion bits of information for recording and the question of the price of such a disc loom large.

Actually, the player would have to be modified as well, since, though the disc itself and the recording and scanning processes are capable of dealing with raw data, the playback electronics are not. The playback electronics seek signals that either are, or seem to be, television signals. Fortunately, it is relatively easy to format data so that it appears as a video signal. All that is necessary is to replace the active picture material with data, while retaining the various synchronizing signals that are used by the playback electronics to identify a video signal. Encoded in this fashion, the data rate might fall to perhaps 3 million bits per second, still allowing more than 10 billion bits on a one-hour disc.

PCM Decoders

If the problems of assembling and selling even that much data seem too great, there are alternative technologies that can make use of data storage on video disc. JVC's AHD and Philips' Compact Disc are specifically designed for the storage and retrieval of digital audio information. But any video disc system can be used for that purpose, in conjunction with the new crop of PCM (pulse code modulation) digital audio decoders now appearing on the market. These units are designed to digitize a stereo audio signal (for essentially perfect quality), format it as video, and record it on an ordinary video tape recorder. In the playback mode, the video signal is reconverted into a raw data stream which, in turn, is converted back into the original stereo audio signal, with no apparent loss from the storage and recovery processes.

Assembling an hour of high fidelity, stereo sound should not present any major problems, nor should the sale of such a disc (assuming a reasonable market size). Best of all, the disc may be played on an unmodified disc player, into an external PCM decoder of the sort that is available now and is in the process of being standardized. Eventually, such decoders, designed specifically for video disc playback-only purposes, might become available at prices well below the present levels (which are in the several thousand dollar range), but the lack of such specialized units need not impair initial market development.

Videotext

Another system of data disguised as video is videotext, also referred to as teletext in its broadcast form. Broadcast teletext is a data stream hidden within ordinary television signals in what is called the vertical blanking interval (the black bar between television frames which becomes visible when a vertical hold circuit is not functioning properly). As it has been implemented in Europe, broadcast teletext can transmit as many as 960 alphanumeric characters hidden within four scanning lines of each frame, which are restored to "pages" of alphanumeric and crude pictorial data by a special decoder. Broadcast teletext is being transmitted in a number of European countries and is being experimented with in the United States as well as in Canada and Japan. A related data access system, often called viewdata, utilizes telephone lines rather than television signals for the data transmission; in many cases, viewdata systems make use of the same decoder for displaying retrieved information on a television set.

Teletext-encoded video information may be recorded on an ordinary entertainment or educational video disc without any effect upon the pictures or sound (although some of the teletext systems being proposed carry information outside of the normal video bandwidth, and these would have to be modified for video disc systems using a restricted bandwidth encoding system). One possible application of such signals might be multilingual captioning. Another might be the transmission of data into a computer while pictorial information is being watched; the computer could, in turn, be programming instructions to modify the television program (as in the action loops referred to earlier). Since teletext is also available with still frames, it might be used to provide descriptive text for a still picture. With some slight modification to utilize more of the vertical blanking interval than the two scanning lines per field traditionally proposed, it might even provide an audio message or music under a still frame, using digital audio.

Except for this last possibility (audio message under a still frame), all of the other uses of teletext in conjunction with video disc systems offer the same benefit as the digital audio PCM decoders. Teletext decoders will proliferate and may be used with unmodified video disc players of any sort as long as the players provide standard television signals.

A Powerful Combination of Technologies

The combination of microprocessor control, picture coding, home computers and data storage techniques can make the video disc player an incredibly powerful, interactive educational tool. Used in conjunction with other data bases (such as continuously updated central computer facilities), it can remove much of the burden from most data transmission systems by acting as an enormous memory for relatively static information. For a computerized lesson in physics, for example, a "Physics Background" video disc could provide the appropriate diagrams instantly, whereas other methods might require hours of data transmission time. Similarly, a "News Background" video disc could provide maps and pictures for news stories fed from a data base. This latter disc is relatively static, but might need to be changed, perhaps biennially, to replace the faces of political representatives with a fresh group.

Clearly, all the possibilities mentioned thus far barely scratch the surface of the video disc's capabilities.

TECHNOLOGIES UNFAVORABLE TO THE VIDEO DISC

However good the video disc appears as an *educational* technology, its future as an *entertainment* technology seems quite bleak. To review, the video disc was developed at a time when the only alternative video storage technology—video tape recording—was expensive, large, difficult to operate and consumed large amounts of tape. That situation rapidly changed.

Video Cassette Systems

In 1971, Sony introduced the U-matic™, 3/4-inch video cassette system. Not only did this system eliminate the nuisance of helical video tape threading, but it also significantly reduced tape consumption from the EIAJ* 1/2-inch rate of 3.75 square inches per second to roughly 2.8 square inches per second. This new video cassette system rapidly took over the educational and industrial markets, killing not only the EIAJ 1/2-inch format, but also a film-based system,

---

*Electronics Industry Association of Japan. See Chapter 2.

developed by CBS, called EVR (electronic video recording). Like most of the video disc systems, EVR's film cartridges could not be recorded by the consumer.

A home video tape cartridge system called Cartrivision was developed by Cartridge Television Inc. (32% owned by Avco Corp.). While the Cartrivision system could record, it was promoted primarily as a playback system for prerecorded movies, recorded with the skip-field technique of dubious quality used by Sony in its Mavica system. Furthermore, throughout most of its brief life, Cartrivision systems were available only as part of large television consoles. Shortly after a stand-alone unit was introduced, Avco withdrew its support and the system failed.

Cartrivision was followed by a system called simply VCR (video cassette recorder), first marketed by Philips in 1972, but aimed primarily at the industrial, rather than the consumer, market. V-Cord, a system jointly developed by Sanyo and Toshiba, followed VCR in 1974, with an indeterminate industrial or home market. However, Matsushita's VX-100 system, introduced in 1975, was unquestionably aimed at the home market: it was introduced at about $750 in Japan, with an hour's tape cassette costing a bit more than $18. Tape consumption in this unit was down to less than 1.75 square inches per second.

Finally, Betamax, a Sony product using some of the principles developed for Mavica, arrived in the United States in February 1976. Its tape consumption was only a bit more than 3/4 of a square inch per second (0.79). Betamax's one-hour capacity was surpassed by JVC's VHS (video home system) capacity of two hours, with a tape consumption of only 0.66 square inches per second, beating even V-Cord II's 0.73. Beta II followed, as did VHS LP, then Beta III and VHS SLP, the present consumption champion at only 0.22 square inches per second. (This is almost a thousandfold improvement over RCA's 1953 prototype video tape recorder, which consumed 180 square inches per second.) In Europe, VCR was followed by VCR-LP and SVR (a Grundig development), and will soon be joined by V2000 (developed by Philips with Grundig), this last to offer an eight-hour capacity. Even VHS in the SLP mode offers only six hours, at the moment, with nine hours expected soon.

Even for a six-hour VHS system, tape cost has fallen to $3 or less, per hour, removing one of the major advantages of video disc systems: significantly lower materials costs. Home video cassette recorders are currently being discounted at prices lower than the only home video disc system currently on the market (the Philips/MCA optical system), thereby removing another possible advantage.

There seems little doubt that video disc systems will offer superior picture quality to those video cassette systems that offer the longest playing time, i.e., consume the least tape. Although the video cassette systems themselves offer improved picture quality as tape consumption increases, consumers are opting for longer-playing units, thereby indicating that the highest quality reproduction is not of great concern. Therefore, it does not seem likely that video discs will be able to enter the prerecorded programming market on the basis of quality of reproduction.

Furthermore, video tape recording technology seems to be moving rapidly toward digital recording techniques. In digital video recording, numbers are recorded, rather than sensitive fluctuations of signal strength. Because it is far easier to retrieve numbers (even from a relatively poor quality tape) than a complex signal, digital video recording offers the potential for a virtually perfect quality storage mechanism. At the moment, not even professional, broadcast units are commercially available. It is reasonable to assume, however, that very shortly after their professional introduction, such units will become available to the home. In fact, digital audio recording units were actually commercially available to the home market before the professional market.

Longitudinal Video Recording

There are only two remaining advantages to video disc systems: the ease and economy of program replication through stamping, molding or contact printing, and the educational advantages of single frame random access (the latest video cassette recorders offer slow motion and high speed search).

Program replication on video tape systems has long been a problem. As described in Chapter 2, creating 1000 copies of an hour-long video tape requires either 1000 hours of recording time on a single recorder (plus rewind and set up time) or one hour on 1000 recorders, or something in between. Attempts at high speed duplication, using thermal or magnetic contact printing, have generally been unsuccessful, and both Ampex and 3M have abandoned their efforts in that area.

There are, however, two new video cassette systems about to emerge which may very well challenge the video disc's superiority in even this area. Both are called LVR (longitudinal video recorder), and both are throwbacks to the earliest video tape technologies of the 1950s—unlike all the other video cassette systems, which rely upon helical video tape technology. Both new systems, being developed separately by Toshiba and BASF, are based on a system proposed in 1965 by the Illinois Institute of Technology.

As in the earliest video tape recorder prototypes, the LVR systems do not use a spinning or rotating head, but simply pass over a fixed head at high speed. This simplifies the mechanical design of the system dramatically, and should allow recorders to be manufactured for significantly lower prices than present home video cassette machines. However, unlike the early prototypes, the LVR units have multiple tracks representing different sections of a video program. When one track is finished, the single head steps down to the next track.

After that, the machines do things differently. BASF's tape is a bit less than 1/3-inch wide and contains 72 tracks, while Toshiba's is 1/2-inch wide and has 300 tracks. BASF fits 600 meters of tape into a cartridge 4 inches square and draws tape from the cartridge at 4 meters per second, reversing direction and stepping to the next track in about 1/10 of a second, for a total playing time of three hours. Toshiba's unit uses an endless loop cartridge (like eight-track tapes), containing 135 meters of tape, which passes the head at 5.5 meters per second, for a playing time of two hours. Since an endless loop cartridge is used, motion does not need to be reversed, and the track change can take place in only 3/100 of a second. Tape consumption on these units is low, but not quite as low as that offered by the longest-playing home video cassette systems now available.

Replication capability

The area in which the LVR systems excel is in their replication capability. Since these are multitrack systems, it is possible to duplicate all of the tracks at once. Even if, for example, it seems impossible to create a single, 1/2-inch, 300-track head, 10 30-track heads or 15 20-track heads could be used. Ideally, duplication would be inside the cassette, using a single head, but even if it has to be done outside the cassette, with multiple heads, the tape can later be loaded into a cassette in the same amount of time it would take to load blank tape in the first place. A 300-track system allows duplication at 1/300 of real time; a 72-track system allows duplication at 1/72 of real time.

What that means is that, without recourse to special magnetic printing techniques, a Toshiba LVR program lasting one hour could be duplicated in exactly 12 seconds! Even the BASF system would allow an hour-long program to be duplicated in just 50 seconds. A 12-second replication time for an hour-long program compares very favorably to the length of time it takes to press a disc. Furthermore, no mastering whatsoever is required. A program just shot can be instantly replicated. There are also few economies of scale (as with the photographic reproduction disc systems). Thus, 10 copies may be created as easily as 1000 copies, although multiple recorders can be used for extremely large runs.

Random access capability

Both systems also offer a random access of sorts. In the Toshiba machine, for example, any 25-second track can be accessed in a maximum time of about eight seconds. That means that access to an individual frame could take a maximum of 33 seconds, assuming there was some way to freeze the frame. (LVR machines have no inherent capability, unlike helical machines, for still frame or variable speed motion.) BASF puts the price of its machine in approximately the same range as a color television set, while Toshiba's range is $500 to $600. Both machines were expected to appear on the market either late in 1980 or early in 1981, but have since been delayed.

The only remaining advantage of video disc systems, and it is a dubious advantage in the consumer market, is the educational capability of intermixed action and still sequences and random access to any particular frame.

A system called RTS (rapid transmission and storage) was developed by Goldmark Communications for the transmission or recording of multiple educational programs, consisting of brief action sequences as well as still frames and continuous audio, on either a single television channel or a single piece of video tape. The number of programs was said to be variable, but 20 was a figure used. Neither this system, nor a system developed by Matsushita in 1973 for presenting color still frames and multichannel audio from an audio cassette, has seen extensive development, though Hitachi demonstrated a system similar to Matsushita's in June 1980.

Frame Storage

There remains, however, another technology, of itself unlikely to compete with video disc systems for the foreseeable future, but, used in conjunction with any video recorder system, a powerful competitor to the educational aspects of the video disc. That technology is frame storage.

Types of systems

Bosch's broadcast video recorder format, referred to as a Type B helical recording system, differs from the Type C helical recording systems in that it cannot provide still frame or slow motion capability. As soon as Bosch recognized that these deficiencies put it at a marketing disadvantage, it developed a frame store accessory to provide slow motion and still frame effects. This is done by recording information from the tape onto a digital, semiconductor memory, which can remember the information even when the tape is stopped. This unit was so successful that the Bosch tape recorder was turned into a random access still frame storage unit: once the frame store has captured a particular frame, the recorder is free to search for the next one. Of course, random access on a tape system takes longer than random access on a disc system (except for LVR, which could actually be shorter than many disc systems).

The Bosch frame storage system costs approximately $15,000. However, a frame store system developed for educational video tape use need not be so high. In 1973 Hughes Aircraft showed a frame store system based on vacuum tube technology (a scan converter tube), rather than on digital, semiconductor memories, and put the price at about $300. Frame stores can also be based on magnetic disc technology (in which case the systems can be relatively inexpensive, since they only need to store 1/30 of a second of video). Hitachi even proposed incorporating a frame store into a television set called "Memory-Matic" in 1974, with a second, smaller screen on which an image would be frozen while action continued on the large screen. Amateur radio operators dealing in a special form of transmission called slow-scan television utilize frame stores, costing less than a few hundred dollars, to reconstruct their images. The point is that, in large quantity, frame stores can be very inexpensive.

A frame store coupled to JVC's VHD video disc system would allow random access to individual frames, instead of to two-frame rotations. Half of a frame store (called a field store) coupled to an 1800 rpm disc system would allow doubling pictorial capability. However, a frame store coupled to JVC's VHS video cassette recorder (rather than its VHD video disc system) can also allow random access to individual frames, through the same frame coding and microprocessor systems used in the educational disc systems. Of course, access time on a VHS system could be excruciatingly long, measured in minutes rather than seconds. On an LVR system, such as Toshiba's, though, a frame store would allow access to any of 216,000 frames (or 432,000 half-resolution fields) within 33 seconds— less time than may be claimed by any video disc system. At current prices, a ham radio quality frame store in conjunction with a Toshiba LVR would also be less expensive than a comparable video disc system. Prices are also expected to drop rapidly. RCA predicts a $100 frame store of high quality by 1985, dropping to $10 by 1990. Finally, as in Bosch's video tape recorder, utilizing a frame store for still pictures allows the recorder (or disc system) to otherwise continue to operate, offering possibilities for audio of almost any length accompanying a still frame.

As already mentioned, frame stores can be comprised of digital, semiconductor memories (like computer memories), vacuum tubes or discs. Once a video signal is digitized (i.e., represented by a series of numbers recognizable to a computer), any sort of digital memory can be used. While semiconductor memories are readily available, other types, such as charge-coupled device memories and bubble memories, might also be used. Since the latter two are relatively newer technologies (the first storing tiny amounts of electrical charge and the second utilizing bubbles of magnetism), their prices may fall at an even greater rate than that of typical semiconductor memories. Furthermore, these other forms of memory, although they have certain shortcomings in computer processing, lend themselves well to the serial nature of television signals. Finally, these memories can be made nonvolatile (that is, requiring no power to continue to store information).

Prices and problems

Therefore, it is possible to conceive of a day when neither video tape nor video discs will be used for television storage systems, and when video will simply be stored in an inexpensive, random access, no-moving-parts, solid-state memory. It is easy to conceive of such a day, but it is impossible to predict when it will occur. If a digital frame store drops to RCA's predicted $10 by 1990, it will still take 108,000 of such stores to hold an hour-long program—that's more than $1 million, a high price to pay for an hour-long program. Even if technological advances improve this prediction a hundredfold or a thousandfold, it is unlikely that consumers would be willing to pay even $1000 for a program contained on a solid-state memory. Eventually prices are certain to come down to the consumer level, but not within the next 10 years.

There is another small problem associated with these tiny memories. Their cost generally goes down as their density increases. As density increases, in turn, the actual number of electrons being used to store a memory cell's state (on or off) shrinks as well. Even in today's densest memories, this number of electrons is low enough to allow stored information to be wiped out by a vagrant cosmic ray. Researchers at IBM and at Yale predict these information losses, called soft errors (because they are not caused by any defective hardware), will occur 3000 times per 1 million hours of operation for the densest memory available today. This is not a figure to worry about, but it might place an absolute limit (in terms of electrons needed to store each bit of information) on the size of the solid-state television memory of the future.

THE NEW VIDEO ENVIRONMENT

Regardless of which medium is used for the delivery of prerecorded programming, video cassettes or video discs, there may turn out to be a limited home market for such programming. As was pointed out previously, the home video market is significantly different from the way it was in the 1960s, when video disc research began.

In cable television, for example, there has been an expansion not only in the number of systems and subscribers (from 640 systems and 650,000 subscribers in 1960 to well over 4000 systems and more than 15 million subscribers in 1980), but also in the channel capacity and services offered by the systems. At the 1980 convention of the National Cable Television Association, all the major manufacturers of cable television equipment were offering the hardware necessary to provide more than 50 television channels on a single cable. Many of these channels are being fed by satellite. There are now three satellites, operated by three different companies, feeding signals ranging from movies and sports to religious programming and news to cable television systems all over the country. One satellite can feed 22 different channels of information, and after the loss of RCA's Satcom III satellite in spring 1980 reduced channel availability, programmers were waiting on line for their chance at satellite transmission. Recent decisions by the Federal Communications Commission, which move toward deregulation of the cable industry, seem certain to result in a significant increase in cable services and subscribers in the near future.

Cable television systems will also be carrying forms of broadcast teletext, including a form which can occupy the entire bandwidth of a television channel, carrying more than 100 times as much information as is possible when data are inserted only in two lines of the vertical blanking interval. A new service, developed jointly by Mattel and General Instrument and called Play Cable™, even provides programming for Mattel's programmable Intellivision™ video game through a special cable adapter. The customer plays the game via cable instead of buying game cartridges. In a similar fashion, consumers can have telephone access to data banks for games and other programming for home computers, instead of purchasing cassettes or discs.

Broadcast television is offering new services as well, including teletext and subscription channels providing movies and sports. Viewdata, the wired form of teletext (accessed via telephone lines) is already in operation in Europe, and is the subject of intense study in the U.S. as well. Comsat, the Congressionally established corporation which represents the U.S. in the International Telecommunications Satellite Organization, would like to initiate direct broadcasting of subscription television signals to the home from satellites by the middle of the 1980s, although its plan faces formidable economic and some technical hurdles, too (there must be no obstruction between viewers and a southern portion of the sky, which eliminates most urban dwellers).

An even more exotic development is a French system called EPEOS, related to the French Antiope videotext system and Didon data transmission system, which adds a further dimension to the video cassette recorder's role in the electronic marketplace. EPEOS would code television programs with both an identifying number, as on video disc frames, and a scrambling system for secure transmission. Someone wishing to record a particular program would purchase a decoding number. This could be different for each subscriber and might be given over the telephone, for credit card billing. That number, entered into an EPEOS decoder, would automatically tune a video cassette recorder to the appropriate channel, descramble the signal, and activate the recorder at the appropriate time. Since the system would operate automatically, it's ideal for transmitting specialized programming in the otherwise dead hours of the night, say 2 a.m. to 6 a.m. Operated on a 50 channel cable television system, this technology would allow 200 hours of programming to be transmitted each night.

Finally, all forms of decentralized information storage may be obviated once a high bandwidth (such as fiber optic), switched, star network (such as the telephone network) becomes available. Unfortunately, like the solid-state, hour-long video memory, such a system is unlikely to develop within the next decade.

In any event, there is a booming market in video programming and information being distributed electronically to the home. Video discs are at best one segment of this market. Some consumers will choose to buy players and accumulate a library of discs; others, overwhelmed by the variety of programming available electronically, will shy away from purchasing a physical carrier for television programming (the disc), and choose a cable or telephone line service to order programs on demand.

SUMMARY

Video disc systems can become excellent educational tools in conjunction with such devices as microprocessors and home computers. Teletext can enhance the video disc's role in this area as well. However, video cassette recorders are rapidly making inroads on the areas of video disc superiority. Raw tape cost is already $3 or less per hour; many machines offer still frame and slow motion; image quality does not seem to be a significant consumer factor; coming LVR machines offer speedy program replication and some measure of random access in a small, inexpensive package; and frame stores will, within five years, allow equivalent random access capability for education and training uses. The amount and diversity of programming transmitted electronically to the home has increased substantially over what was available when most video disc systems were being developed. The effect this will have on discs, which are physical carriers for electronic programming, is unknown. However, video disc players clearly must be seen as one technology, not *the* technology, for delivering programming and information to users.

9

Conclusions

by Efrem Sigel

RCA has stated publicly that its investment in the video disc will exceed the $130 million it sank into color television. When the tens of millions expended by Philips, MCA, Matsushita, Telefunken and other major companies are added in, the worldwide total to develop the disc may exceed $1 billion. Most of this money has been spent on consumer players, although significant sums have also been devoted to educational and industrial applications. Surely this level of investment insures that the video disc has a future. Or does it?

Dollars, yen and marks by themselves are no guarantee that video discs will find a place in the environment of communications devices. Only clear technical advantages at a price consumers can afford will provide such a guarantee.

As Chapter 8 showed, there is no lack of competing technologies for storing and playing back television images. Video tape recording has made steady, sometimes astonishing progress, as engineers have figured out how to pack more and more information onto the same —or narrower—widths of tape. Miniaturization and the application of solid-state electronics have produced smaller, less expensive and highly reliable playback devices. In 1980 tape prices seemed poised for a drop. Thus, it's by no means certain that plastic discs will ultimately be dramatically less costly than reels of video tape coated with metal oxides.

Electronic frame stores, videotext and other video technologies also threaten to impinge on the video disc and its applications. The effect of personal computers is yet to be gauged. In one important respect, the personal computer is an ally, since the disc offers another means of storing information for these devices.

In another sense, computers are competitors for consumer pur-
chases. If there is one iron law of economics, it is that customers cannot spend more money than they have—at least, not for long—no
matter how attractive the items they'd like to buy. This rule should
serve as a reminder that the proponents of any new technology are
not the ultimate determinants of its success. Buyers are.

There are two kinds of customers: users at home, and institutional
buyers in business, education, government, medicine. These distinct
markets will be discussed in turn.

THE CONSUMER MARKET

One factor in assessing the consumer market for video discs is the
burgeoning cable television industry. Because cable TV in mid-1981
reached 19.6 million homes, with almost 10 million of them taking
pay TV channels (see Chapter 3), consumers already had much more
diversity in video entertainment and information than was true even
three years before. FCC rulings in the summer of 1980, ending restrictions on the number of signals and type of programming a cable
system may offer, can only add to this diversity. The price of a disc
player, and of the discs to go with it, must be weighed against the
monthly cost of subscribing to cable and pay TV, with their proliferation of movie, sports and information channels.

Also, in the consumer market, disc players will stand or fall mainly
on their attractiveness vis-a-vis video cassette recorders. As Table 9.1
shows, VCRs have a clear advantage in several of the features that
have thus far been important to consumers—the ability to record off
the air and to make their own programs. Both systems can play back
programs, while discs offer advantages in stereo sound, random access, showing of still pictures and interactive programs. As for price,
no one can say for sure. Disc proponents have been insisting for years
that their system would offer low prices for both players and discs,
but this claim has yet to be substantiated. Meanwhile, discounting
and the marketing of stripped-down models has effectively reduced
VCR prices to under $800 in many stores, and a price of $600 seemed
well within reach by the end of 1980.

The one area in which discs have a clear-cut lead is in replication
cost: the raw material and the stamping process both lend themselves
to mass manufacturing of discs that is far more efficient than the

Table 9.1 Comparative Features of Tape vs. Disc Systems

	Record Off the Air	Play Pre-recorded Programs	Record Own Programs (with camera)	Random Frame Selection	Still Pictures	Stereo Sound	Inter-active Programs	Low-cost Players	Low-cost Programs
Tape	✓	✓	✓					✓	
Disc		✓		✓	✓	✓	✓	✓	✓

Table 9.2 Value of Disc Features to Consumers vs. Institutional Users

	Play Pre-recorded Programs	Random Frame Selection	Still Pictures	Stereo Sound	Inter-active Programs	Low-cost Players	Low-cost Programs
Consumers	✓			✓		✓	✓
Institutional users		✓	✓		✓		

comparable process in tape duplication. Yet even if the cost of reproducing a one-hour disc in quantity turns out to be $1, vs. perhaps $5
for a tape, this would not give discs an insurmountable advantage.
Since most of the cost of a disc or tape to the consumer will consist of
program royalty fees, distribution costs and profit, it would be possible for program distributors to pass on the physical duplication of the
tape at cost, leading to a price comparison of say, $20 for a one-hour
tape vs. $15 for a one-hour disc. Since the tape is reusable, and the
player is also a recorder, the consumer may well feel the $5 difference
is worth it.

By analyzing the implications of Table 9.1, it becomes apparent
that the technical features at which disc systems excel are those with
which consumers have little experience, and no knowledge of how to
use. Take random access: feature films are not made to be advanced
one frame at a time or to be searched for a single episode. Also, the
ability to show interactive programs (those that challenge the viewer
to answer questions and to show mastery of a subject) scarcely exists
today outside of the experimental production centers described in
Chapter 4 at Nebraska, Utah State and Brigham Young universities.
Low disc replication costs, therefore, will be meaningless if it winds
up costing hundreds of thousands of dollars to develop each hour of
interactive programming.

On the other hand, those features unique to VCRs are demonstra-
bly important to consumers: surveys have shown that the ability to
record broadcast TV programs is the most important reason why
consumers buy the equipment.

Another way of analyzing the disc features is to cast them in terms
of their appeal to consumers and to institutions. This is the approach
taken in Table 9.2. The features most important to consumers are the
ability to play prerecorded programs, along with stereo sound and
low prices for both players and discs. Institutions, on the other hand,
while also interested in playing back programs, are far more able to
make use of such innovative disc capabilities as random frame selection, the showing of still pictures and the availability of interactive
programs.

And although business and educational organizations would naturally like to buy disc players at the lowest possible price, they will not

avoid purchasing them simply because the price happens to be $2000 instead of $500. Certainly the sales of industrial VCRs costing close to $2000, and of industrial disc players costing close to $3000, indicate that the capabilities of a piece of equipment are more important than a bare-bones price to these users.

This brings us, then, to the question of the total amount of money that consumers and institutions have available to buy disc players and programs. If one or the other market had a large lead in revenues available for this purpose, it should have an undeniable effect on the nature of programming developed, the features stressed by hardware manufacturers and the marketing approaches used.

In the consumer market, the relevant dollar amount is what consumers spend on leisure time activities, and specifically, for records, books, newspapers, movies and electronic equipment. The total is very large—around $25 billion for 1979—and suggests that there is no lack of consumer funds that might go for buying players and discs. (See Table 9.3.)

Table 9.3 Sales of Publications, Audiovisual Programs and Equipment to Consumers, 1979

Category	Manufacturer or Publisher Sales (millions)
Consumer electronics equipment (TVs, radios, VCRs, phonographs, etc.)	$ 9,500
Records and tapes	2,450
Cable TV subscription revenues	1,450
Motion picture box office receipts	2,950
Newspaper circulation revenues	4,200
General magazine circulation revenues	1,900
Book sales (general trade, paperback and mail order books only)	2,500
Total	$24,950

Source: 1980 U.S. Industrial Outlook, and Knowledge Industry Publications, Inc. projections for 1979 based on 1977 Census of Manufactures.

THE INSTITUTIONAL MARKET

Set against this consumer market is the business market for information machines and related services, described in Table 9.4 It includes spending for computer equipment and other office and photo-
copying equipment, as well as audiovisual equipment like broadcast
TV hardware. Another category of institutional spending involves
salaries and supplies for people engaged in training, media production, industrial films, etc. as well as in computer departments. Obviously this encompasses a wide range of activities, and the resulting
total of $66 billion in expenditures must be regarded not as the
market for video discs, but as a rough measure of the staggering annual outlays that business makes for communications activities.

The broader point is this: the largest opportunities for any new
technology involve getting users to substitute capital for labor. If corporations will shift even a tiny fraction of the billions being spent on
salaries of those engaged in training, or data processing, or audio-

**Table 9.4 Institutional Spending on Selected Communications/
Information Machines and Related Activities, 1979**

Category	Expenditures (millions)
Equipment	
Computers	$21,100
Calculating and accounting machines	1,000
Office photocopiers	3,700
Micrographic equipment	1,000
Broadcast radio/TV equipment	780
Subtotal, equipment	27,580
Activity	
Computer department salaries and supplies	27,700
Industrial training	10,000
Nonbroadcast TV salaries and services	750
Subtotal, salaries and services	38,450
Total, communications/information services	$66,030

Source: 1980 U.S. Industrial Outlook, Knowledge Industry Publications, Inc.,
and International Data Corp.

visual production, into the purchase of disc players and programs, then the market will be large indeed.

In the consumer market, these considerations don't apply. Consumers purchase TVs or video disc players not as labor-saving devices, but because they make life more pleasant. Manufacturers are spared the chore of demonstrating that the equipment does a task more efficiently—e.g., it is not necessary to show consumers that the video disc is a better teaching medium than a sound filmstrip projector or a VCR, in order to persuade them to purchase it. But manufacturers do have to find exactly the right blend of equipment features and programming to appeal to always changing, and naturally diverse, consumer tastes.

STANDARDIZATION

In both consumer and institutional markets, lack of standardization poses an obstacle to widespread adoption of video disc players. But standardization cannot be pursued blindly. Although the consumer market will influence the institutional market, the reverse is not true. A disc player that is successful in the consumer market will be low-priced, and therefore attractive to companies and schools. But a high-priced disc player can be purchased by educators and trainers without awakening any desire on the part of consumers to own it, in much the same way that photocopiers have become staples of modern business life while scarcely appearing in homes.

It is entirely possible that two or even three different disc standards will emerge: 1) a low-priced player without any interactive features, for use in homes and in those institutions that only need to play back traditional types of programs; 2) a more sophisticated player with interactive features, for use in training and communications in business and education; and possibly 3) a video disc for data storage, using optical playback, and perhaps, as research into this field progresses, eventually permitting some updating of information on the disc. (*Scientific American* mentions this possibility in an August 1980 article on "Disk Storage Technology," by Robert A. White.) Because of the different features required for each of these uses, it is misleading and ultimately futile to talk about "the" video disc as if every purchaser had to have the identical device. After all, sports cars, sedans, family campers and giant tractor trailers all have their uses, and no one insists that everyone should drive the same four-wheel vehicle for the sake of standardization.

FUTURE OF VIDEO DISCS

Thus, we come back to considering the video disc as a family of visual display devices, rather than as a single item, and to understand its role we must put it into the context of our evolving communications technology. The years from 1950 to 1980 saw the rise of television from a novelty to the most powerful communications medium of our era, with growing impact on politics, business, education, even religion and family life. Never has the label "mass media" been more apt than in describing network television in the U.S., or its counterpart in Europe or Japan: the ability of a single station to command the simultaneous attention of 10, 20, even 40 million homes dwarfs anything comparable in human history.

But cable television, video cassette recorders and now video disc players have altered forever the nature of video communications. These latter two devices are under the control not of a central broadcasting authority but of a single viewer, whether at home or in an office or classroom. What he chooses to watch depends on his own interests and vocational needs; whether he chooses to watch may well depend on the skill of disc manufacturers and suppliers in divining those needs.

There is no gainsaying the many fits and starts that have plagued the development of video discs. Economic issues, lack of standardization and competition from different technologies will all affect the future of the disc. Nevertheless, video discs have enough unique features that persistent experimentation seems bound to uncover those that most appeal to users, at an affordable price. When that happens, video discs will take their place as one of an array of fascinating accessories to the television set.

Appendix: Organizations Involved with the Video Disc as of October 1980

EQUIPMENT MANUFACTURERS/DISTRIBUTORS

ARDEV Co., Inc.
(the video disc division of McDonnell Douglas Electronics Co., a subsidiary of McDonnell Douglas Corp.)
1057 East Meadow Circle
Palo Alto, CA 94303
Will manufacture ARDEV interactive photographic film video disc system (industrial/educational).

DiscoVision Associates
(Joint venture of IBM and MCA, Inc.)
100 Universal City Plaza
Universal City, CA 91608
Distributes industrial video disc player (Universal-Pioneer), using Philips laser optical system.

General Corp.
1116 Suenaga Takatsu-ku
Kawasaki, Kanagawa 213, Japan
Manufactures and distributes TeD video disc systems (in Japan).

General Electric Co.
3135 Easton Turnpike
Fairfield, CT 06431
Will manufacture video disc players using JVC VHD capacitance system.

IBM Corp.
(See DiscoVision Associates)

Magnavox Consumer Electronics Co.
(Subs. of N.A. Philips)
Interstate 40 & Straw Plains Pike
P.O. Box 6950
Knoxville, TN 37914
Manufactures and distributes consumer (Magnavision) and industrial video disc players using Philips laser optical system.

Matsushita Electric Corp. of America
1 Panasonic Way
Secaucus, NJ 07094
Will distribute video disc players using JVC VHD capacitance system.

Matsushita Electric Industrial Co., Ltd.
1006 Kadoma
Osaka, 571, Japan
Will manufacture video disc players using JVC VHD capacitance system.

N.A. Philips
(see Magnavox Consumer Electronics Co.)
In October 1980, Philips signed an agreement (pending Justice Dept. approval) to acquire GTE/Sylvania, which is expected to enter the video disc market.

N.V. Philips Gloelampenfab-riken
Audio/Video Department
Eindhoven, The Netherlands
*Manufactures and distributes
video disc players using Philips
laser optical system.*

Panasonic Video Systems Div.
(Div. of Matsushita Electric
Corp. of America)
1 Panasonic Way
Secaucus, NJ 07094
*Will distribute video disc players
using JVC VHD capacitance
system.*

J.C. Penney
1301 Ave. of the Americas
New York, NY 10019
*Will distribute consumer video
disc players, using RCA capacitance system, under J.C. Penney
label.*

Quasar Electronics Co.
(Div. of Matsushita Electric
Corp. of America)
9401 Grand Ave.
Franklin Park, IL 60131
*Will distribute video disc players
using JVC VHD capacitance
system.*

RCA Consumer Electronics Div.
600 N. Sherman Dr.
Indianapolis, IN 46201
*Manufactures and distributes
video disc players (SelectaVision)
using RCA capacitance system.*

Sears Roebuck
Sears Tower
Chicago, IL 60684
*Will distribute consumer video
disc players, using RCA capaci-
tance system, under Sears label.*

Sony Corp.
7-35 Kitashinagawa 6-chome,
Shinagawa-ku
Tokyo, 141, Japan
*Manufactures industrial video
disc players (at Atsugi Plant)
using Philips laser optical
system.*

Sony Corp. of America
(Subs. of Sony Corp.)
9 W. 57th St.
New York, NY 10019
*Distributes industrial video disc
players using Philips laser optical system.*

**Telefunken Television and
Radio Corp.**
Goettinger Chaussee 76
3000, Hannover 91
West Germany
*Manufactures industrial/institu-
tional video disc players using
the TeD system.*

Thomson-CSF Broadcast, Inc.
37 Brownhouse Rd.
Stamford, CT 06902
*Will manufacture and distribute
industrial video disc player using
Thomson-CSF laser optical
system.*

U.S. Pioneer Electronics Corp.
(Subs. of Pioneer Electronic
Corp.)
85 Oxford Dr.
Moonachie, NJ 07074
Distributes consumer (Laserdisc)
and industrial (Universal
Pioneer) video disc players using
Philips laser optical system.

U.S. JVC Corp.
(Subs. of Victor Co. of Japan,
an affiliate of Matsushita)
41 Slater Dr.
Elmwood, NJ 07407
Will distribute video disc players
using JVC VHD capacitance
system.

Universal-Pioneer Corp.
(Joint venture of Pioneer Elec-
tronic Corp. and DiscoVision
Associates)
4-1 1-chome
Meguro, Meguro-ku
Tokyo, 153, Japan
Manufactures video disc players
using Philips laser optical
system.

Victor Co. of Japan
(An affiliate of Matsushita)
1, 4-chome Nihonbishi-honcho
Chuo-ku
Tokyo, 103, Japan
Will manufacture video disc
players using JVC VHD capaci-
tance system.

Zenith Radio Corp.
1000 Milwaukee Ave.
Glenview, IL 60025
Will distribute consumer video
disc player, using RCA capaci-
tance system, under Zenith
label. Has announced plans to
manufacture players at an
unspecified future date.

PROGRAMMING PRODUCERS/DISTRIBUTORS

ABC Video Enterprises
1330 Ave. of the Americas
New York, NY 10019
Produces programming (ABC
School Discs) for video disc
players using Philips laser opti-
cal system.

ARDEV Co., Inc.
(see Equipment Manufacturers/
Distributors for address)
Will produce programming for
video disc players using ARDEV
system.

*Brigham Young University
David O. McKay Institute of
Education
W-160 STAD
Provo, UT 84602
Produces interactive instructional programming for video disc players using laser optical format.

*Caravatt Communications
551 Fifth Ave.
New York, NY 10017
Produces programming for video disc players using Philips laser optical system.

CBS Video Enterprises
51 W. 52nd St.
New York, NY 10019
Distributes programming for video disc players using RCA capacitance system.

MCA DiscoVision Inc.
100 Universal City Plaza
Universal City, CA 91608
Acquires and distributes programming for video disc players using Philips laser optical system.

*Nebraska Video Disc Design/ Production Group
P.O. Box 83111
Lincoln, NE 68501
Designs, produces and provides consultation for interactive video disc programming in all laser optical formats.

Optical Programming Associates
(Joint venture of Philips, Universal Pioneer and MCA, Inc.)
c/o Magnavox Productions, Inc.
100 E. 42nd St.
New York, NY 10017
Acquires, produces and distributes programming for video disc players using the Philips laser optical system.

Pioneer Artists, Inc.
(Subs. of U.S. Pioneer Electronics Corp.)
85 Oxford Dr.
Moonachie, NJ 07074
Acquires, produces and distributes programming (mainly music) for video disc players using Philips laser optical system.

*Sandy Corp.
16025 Northland Dr.
Southfield, MI 48075
Designs, produces and provides consultation for video disc programming in all current formats, specializing in training communications systems for business and industry.

Utah State University
Videodisc Innovations Project
Logan, UT 84332
Produces interactive programming for video disc players using laser optical format.

---

*Have announced plans to produce programming in all future video disc formats as well.

WICAT, Inc.
1160 S. State St.
Orem, UT 84057
Produces interactive program-
ming for video disc players
using laser optical format.

The following companies have licensed titles for consumer video disc release:

ABC, Inc.
Avco Embassy Pictures
CBS, Inc.
Columbia Pictures
Corp. for Entertainment
  & Learning
Filmverhuurkantoor de Dam
  B.V.
ITC Entertainment Ltd.
The Jackson Co.
MCA-Universal Pictures
MGM, Inc.

NBC, Inc.
NFL Films, Inc.
Paramount Pictures Corp.
Rank Film Distributors
Time-Life Films
Twentieth Century-Fox Film
  Corp.
United Artists Corp.
United Features Syndicate
Viacom Enterprises
Walt Disney Productions
Warner Bros.

DISC RESEARCH/DISTRIBUTION/REPLICATION

ARDEV Co., Inc.
(see Equipment Manufacturers/
Distributors for address)
Will replicate and distribute for
players using ARDEV system.

CBS, Inc.
51 W. 52nd St.
New York, NY 10019
Disc pressing plant under con-
struction for replication of discs
for players using RCA capaci-
tance system.

DiscoVision Associates
(see Equipment Manufacturers/
Distributors for address)
Replication for players using
Philips laser optical system.

IBM Corp.
Armonk, NY 10504
Research in disc development
technology for players using
Philips laser optical system.

3M
3M Center
St. Paul, MN 55101
Replication for players using
Philips laser optical and
Thomson-CSF optical systems.

Producers Color Service
16210 Meyers Rd.
Detroit, MI 48235
Will replicate discs for players
using Philips laser optical
system.

RCA Consumer Electronics Div.
(see Equipment Manufacturers/
Distributors for address)
Replication and distribution for
players using RCA capacitance
system.

Sony Corp. of America
(see Equipment Manufacturers/
Distributors for address)
Disc replication for players
using Philips laser optical
system.

TelDec
Finckenstein Allee 38
1000 Berlin 45
West Germany
Disc replication for players
using TeD system.

Thorn EMI, Ltd.
Upper Saint Martin's Lane
London, WC2H 9ED, England
Will replicate discs for players
using JVC VHD capacitance
system.

**Xerox Information Products
Group**
800 Long Ridge Rd.
Stamford, CT 06904
Will develop optical disc technol-
ogy for data processing appli-
cations with Thomson-CSF.

Index

About the Authors

Efrem Sigel is editor in chief of Knowledge Industry Publications, Inc. He is editor and co-author of *Videotext: The Coming Revolution in Home/Office Information Retrieval.* He is also author of *The Kermanshah Transfer*, a novel, and *Crisis: The Taxpayer Revolt and Your Kids' Schools.* A graduate of Harvard College and Harvard Business School, he has been a teacher and a Peace Corps volunteer.

Mark Schubin is a technological consultant to various organizations including Lincoln Center for the Performing Arts. He is technical editor of both *Videography* and *Home Video* magazines, and has published nearly 100 articles on television technology. He was recently awarded his second Emmy for technological achievements in broadcasting. His professional memberships include the Royal Television Society, the Society of Motion Picture and Television Engineers and the Institute of Electrical and Electronics Engineers. He holds a B.E.Ch.E. from Stevens Institute of Technology.

Paul F. Merrill is professor of instructional science at Brigham Young University. He previously served as director of course development at the University of Mid-America and as associate professor of instructional design at Florida State University. A graduate of Brigham Young University, he holds a Ph.D. from the University of Texas at Austin. He is currently writing a textbook on task and content analysis.

Kenneth S. Christie is unit director of the Nebraska Videodisc Design/Production Group at KUON-TV, University of Nebraska-Lincoln. Previous positions were as music director and manager of radio station KRNU-FM, and as a news reporter/photographer for WOWT-TV. He holds a degree in broadcast journalism and sociology from the University of Nebraska-Lincoln.

John Rusche is vice president of production for Sandy Corp. He has been in charge of Sandy Corp.'s video disc production effort since the project's inception. He is a manager in the Detroit section of the Society of Motion Picture and Television Engineers and a member of the Society's study group on video disc systems. He holds a BA in communications and an MBA from Wayne State University.

Alan Horder is research officer with the National Reprographic Centre for documentation at the Hatfield Polytechnic, England. He is author of *Video Discs—Their Applications to Information Storage and Retrieval.*

Other Titles from Knowledge Industry Publications

Videotext: The Coming Revolution in Home/Office Information Retrieval
edited by Efrem Sigel
154 pages   hardcover $24.95

The Executive's Guide to TV and Radio Appearances
by Michael Bland
144 pages   hardcover $14.95

Practical Video: The Manager's Guide to Applications
by John A. Bunyan, James C. Crimmins and N. Kyri Watson
203 pages   softcover $17.95

The Video Register, 1980-81 edition
250 pages (approx.)   softcover $34.95

Television and Management: The Manager's Guide to Video
by John A. Bunyan and James C. Crimmins
154 pages   hardcover $17.95

Video in the Classroom: A Guide to Creative Television
by Don Kaplan
161 pages   softcover $24.50

Viewdata and Videotext, 1980-81: A Worldwide Report
Transcript of international conference
624 pages   softcover $75.00

Video in the 80s: Emerging Uses for Television in Business, Education, Medicine and Government
by Paula Dranov, Louise Moore and Adrienne Hickey
186 pages   hardcover $34.95

Video User's Handbook
by Peter Utz
410 pages   hardcover $19.95

The Cable/Broadband Communications Book, Volume 2, 1980-81
edited by Mary Louise Hollowell
300 pages (approx.)   hardcover $29.95

Available from Knowledge Industry Publications, Inc., 2 Corporate Park Drive, White Plains, NY 10604.